Introduction to Robots and Robotics
机器人专业英语

王耀军　主编

北京航空航天大学出版社

内 容 简 介

本书针对工业机器人技术技能人才培养的需求，以机器人相关专业的英语能力培养为目标，旨在使读者掌握工业机器人专业英语知识，以及机器人技术应用现状和发展趋势。

全书正文部分由 11 个单元组成，涵盖了机器人主要概念和技术内容，包括基本定义和知识、机器人的类型和应用、机器人的主要结构、机器人的动力驱动、机器人的信号检测、机器人的控制与编程、机器人末端执行器、机器人运动学和动力学基础、机器人性能指标、机器人安全和伦理、机器人发展趋势和展望；附录部分由 7 个单元组成，包括科技英语、职业规划和简历撰写、机器人从业人员职业素养、文献检索、讨论和思考、机器人大观园、生词释义。

本书突出了材料的新颖性和资料的趣味性，兼具内容的拓展性，既注重专业能力培养，又融入职业核心能力培育，可作为高职高专机器人相关专业的专业英语教材，也可作为工程技术人员的参考资料和培训用书，亦可作为科普读物供广大关注机器人发展的读者使用。

图书在版编目(CIP)数据

Introduction to Robots and Robotics 机器人专业英语 / 王耀军主编. －－北京：北京航空航天大学出版社，2021.2

ISBN 978-7-5124-3169-0

Ⅰ. ①I… Ⅱ. ①王… Ⅲ. ①工业机器人－英语－高等职业教育－教材 Ⅳ. ①TP242.2

中国版本图书馆 CIP 数据核字(2020)第 263558 号

版权所有，侵权必究。

Introduction to Robots and Robotics
机器人专业英语

王耀军　主编

策划编辑　周世婷　责任编辑　冯　颖

*

北京航空航天大学出版社出版发行

北京市海淀区学院路 37 号（邮编 100191）　http://www.buaapress.com.cn
发行部电话：(010)82317024　传真：(010)82328026
读者信箱：goodtextbook@126.com　邮购电话：(010)82316936
北京建宏印刷有限公司印装　各地书店经销

*

开本：787×1 092　1/16　印张：8.75　字数：224 千字
2021 年 2 月第 1 版　2025 年 1 月第 4 次印刷
ISBN 978-7-5124-3169-0　定价：29.00 元

若本书有倒页、脱页、缺页等印装质量问题，请与本社发行部联系调换。联系电话：(010)82317024

前　言

得益于国家产业升级换代，各院校纷纷开设机器人专业（方向），但由于起步较晚，机器人专业英语一直缺乏定位技术技能人才培养的教科书。笔者从事多年一线教学工作，深感缺少专科层次、适合机器人专业（方向）、内容新颖且重点突出的专业英语教材。

因此，笔者在加拿大麦吉尔大学智能机器研究所（Centre for Intelligent Machines，McGill University）跟随机器人学科权威 Jorge Angeles 教授学习期间，进行了一些教学探索和思考，并针对国内机器人专业（方向）的需求和定位，结合国际主流机器人专业的教学包，编写了本教材。虽在编写过程中屡屡受挫，但得到了教授的鼓励、CIM 实验室同仁的支持和浙江机电职业技术学院机电一体化以及自动化专业各位老师的帮助，本教材最终得以成形。

本书立足于机器人专业及学科的技术体系，围绕机器人专业学生的培养目标，突出"内容新颖、语言地道、简明实用、适合教学"的特色，具体表现在以下几方面：

（1）包含机器人专业学习的主要内容，而且选材新颖、饶有趣味，通过学习本课程，既可掌握机器人专业英语知识，还可全面了解本专业的技术体系；

（2）内容取材和编译经过以英语为母语的业内专家学者编辑、修改，避免"中国式英语"现象，呈现地道的机器人学术英语；

（3）编排结构简明，内容实用，前半部分为机器人专业涉及的专业内容，后半部分还给出了本专业学生升学或就业所需的科技英语、职业素养、职业规划与英文简历、文献检索、机器人资源大全等内容；

（4）充分考虑到教与学的需求，以模块化方式编排各章，章节之间有机衔接，章节内部又自成一体，便于学生自学和拓展，教师亦可根据学时灵活调整教学内容。

限于编者水平，书中难免有错误和不妥之处，恳请读者批评指正。

编　者
2021 年 1 月

Contents

Chapter 1　Introduction: robots and robotics ··· 1

  1.1　What is a robot? ·· 1

  1.2　What can robots do (or what are they supposed to do) ················ 3

  1.3　The elements of an industrial robot ·· 5

Chapter 2　World of robots ·· 9

  2.1　Assembly-line robots: robots for making things ···························· 9

  2.2　Portering robots: robots that fetch and carry ······························ 11

  2.3　Tele-operated robots: keeping the human in the loop ················ 13

  2.4　Robots for education ·· 20

  2.5　Other Smart robots ··· 21

Chapter 3　Actuators for robots ·· 24

  3.1　Pneumatic ·· 24

  3.2　Hydraulic ·· 25

  3.3　Electric ··· 26

  3.4　Other electromechanical actuators ·· 26

  3.5　Mechanical transmission ·· 27

Chapter 4　Sensing for robots ·· 28

  4.1　Sensing in general ··· 28

  4.2　Internal sensing ··· 30

  4.3　External sensing ·· 31

Chapter 5　Controlling and programming for robots ······································ 34

  5.1　Methods of operation of industrial robots ·································· 34

  5.2　Various coordinate systems ·· 35

  5.3　Methods of teaching and programming ····································· 37

  5.4　Programming languages for industrial robots ···························· 39

Chapter 6 Wrist, hand, and gripper … 41

6.1 Introduction … 41
6.2 Wrist … 41
6.3 Gripper … 42
6.4 Future robotic hands … 43

Chapter 7 Robot kinematics and dynamics … 45

7.1 Robot Kinematics … 45
7.2 Dynamics … 47

Chapter 8 Performance specification of industrial robots … 49

8.1 Physical characteristics of robots … 49
8.2 Geometric configuration … 50
8.3 Positioning accuracy and repeatability … 51
8.4 Angular accuracy and repeatability … 52
8.5 Speed and acceleration accuracy … 53
8.6 Spatial specifications: working volume, swept area, reach … 53
8.7 Payload … 54
8.8 Vibration … 54
8.9 Miscellaneous specifications … 54

Chapter 9 Industrial robots: a case study … 55

9.1 Application of industrial robots … 55
9.2 A case study: Fanuc M-10iA/12 robot arm … 57
9.3 Picking and packing workstation … 59
9.4 Outlook … 59

Chapter 10 Safety and ethics for robots … 60

10.1 Safety … 60
10.2 Robot ethics … 64

Chapter 11 Robotic futures … 67

11.1 Future breakthrough on robotics … 67
11.2 Autonomous robot planetary scientist … 67

11.3 A swarm of medical micro-robots ········· 69
11.4 A humanoid robot companion ········· 70

Appendix A English for Science and Technology(EST) ········· 72

A.1 Characteristics of grammar and sentences of EST ········· 72
    A.1.1 Commonly used: passive verb ········· 72
    A.1.2 Commonly used: the simple present tense ········· 73
    A.1.3 Commonly used: long sentence ········· 73
A.2 Characteristics of words ········· 75
    A.2.1 Specialized general words ········· 75
    A.2.2 Prefix and suffix of words ········· 75
    A.2.3 Abbreviation and acronym ········· 76
    A.2.4 Combined words ········· 76

Appendix B How to plan your career and create a resume ········· 77

B.1 How to plan your career? ········· 77
B.2 How to create a resume? ········· 79
    B.2.1 Structuring your resume ········· 79
    B.2.2 Making your content shine ········· 80
    B.2.3 Finalizing your resume ········· 81
    B.2.4 Resume template ········· 81

Appendix C Code of ethics for robotics engineers ········· 85

C.1 Preamble ········· 85
C.2 Principles ········· 85
C.3 Conclusion ········· 87

Appendix D Information retrieval ········· 88

D.1 Information retrieval, more important than you thought ········· 88
    D.1.1 What is information retrieval ········· 88
    D.1.2 What can information retrieval do? ········· 88
    D.1.3 How to do the literature searching? ········· 90
D.2 Retrieval of Chinese database: CNKI ········· 90
D.3 Retrieval of English database: ISI Web of Knowledge ········· 97
D.4 Other resources ········· 101

D.4.1　Bing Academic ·· 101

  D.4.2　Google Scholar ··· 102

  D.4.3　New ways of information retrieval ··· 104

Appendix E　Tell & Show ·· 107

  E.1　Chapter 1 ··· 107

  E.2　Chapter 2 ··· 107

  E.3　Chapter 3 ··· 107

  E.4　Chapter 4 ··· 108

  E.5　Chapter 5 ··· 108

  E.6　Chapter 6 ··· 108

  E.7　Chapter 7 ··· 108

  E.8　Chapter 8 ··· 108

  E.9　Chapter 9 ··· 109

  E.10　Chapter 10 ·· 109

  E.11　Chapter 11 ·· 109

Appendix F　World of robotics ·· 110

  F.1　Classic books on robots and robotics ·· 110

  F.2　Journals and magazines on robots and robotics ····························· 111

  F.3　Websites on robots and robotics ··· 111

  F.4　Robotics open courses ··· 112

  F.5　Other resources ··· 113

Appendix G　Words and expressions ·· 114

Reference ·· 130

Acknowledgement ·· 131

Chapter 1

Introduction: robots and robotics

什么是机器人？它是机器还是人？它是由哪些部分组成的？如何学习和应用机器人？同学们，欢迎来到机器人的世界，让我们一起来探索机器人，了解机器人的现状和发展趋向，解开这诸多疑问。

1.1 What is a robot?

The word *robot* was used for the first time in a play called "Rossum's Universal Robots" by the Czech dramatist Karl Kapek. The word *Robot* means forced labor or serf (peasant). In the 1920s science fiction play, which portrayed robots as intelligent machines serving their human makers, the plot ended dramatically. Robots took over the world and destroyed humanity. This scenario is far from reality. Later, Isaac Asimov coined and popularized the term *robotics* through many science-fiction novels and short stories. Unlike earlier robots in science fiction, robots do not threaten humans since Asimov invented the well-known *Three Laws of Robotics*[1], which will be introduced in Section 10.2. To say that the division between human and robot is perhaps not as significant as that between intelligence and non-intelligence is not implausible.

The first industrial robot, the PUMA[2], was built by Unimation in 1961. Before going on, have a look at the popular *droids and humanoid robots*[3] shown in Fig. 1.1.

Today industrial robots and robotic systems are key components of automation. More than two million industrial robots are operating in the factories all over the world:
- Improving quality of work for employees.
- Increasing production output rates.

---

① 机器人三大定律，将在安全与伦理章节(Chapter 10)详细介绍。
② Programmable Universal Manipulator Arm，即可编程通用操作臂机器人。
③ 机器人和类人机器人。

Fig. 1.1 Pop-culture droids and humanoid robots

- Improving product quatlity and consistency.
- Increasing flexibility in product manufacturing.
- Reducing operating costs.

Therefore defining robotics concisely is a difficult task, as it is a subject without sharp boundaries: at various points on its periphery it merges into fields such as intelligence, automation and remote control. Moreover, the boundaries of robotics are not only vague but also shifting. Robots are evolving quickly, as with our ideas given to them, so that we expect more and more intelligence from machines. Given this situation, it is unwise to insist on a rigid definition of "robot" or "robotics", but the following list of characteristics seems to be essential for a true robot:

- A robot must be produced by manufacturing rather than by biology.
- It must be able to move physical objects or be mobile itself.
- It must be a power or force source or amplifier.
- It must be capable of some sustained action without intervention by an external agent.
- It must be able to modify its behavior in response to sensed properties of its environment, and therefore must be equipped with sensors.

A less formal view of a robot is that it is a machine possessing functional arms or legs, or else is a driverless vehicle. Other definition emphasize intelligence, by which is meant the human-like ability to perform a variety of incompletely specified tasks involving *perception*④ and *decision making*⑤.

The official definition of robot issued by the Robot Institute of America (RIA) is: *A robot is a reprogrammable and multifunctional manipulator, devised for the transport of materials, parts, tools or specialized systems, with varied and programmed movements, with the aim of carrying out varied tasks.*

Robotics is the science of robots, and humans working in this area are thus being called roboticists. Now you are supposed to have a clearer understanding of a robot, let us move on to learn what robots can do and why.

1.2 What can robots do (or what are they supposed to do)

Traditionally, robots are applied anywhere one of the 3Ds exists: in any job which is too *Dangerous*, *Dull*, and/or *Dirty* for a human to perform.

1. Jobs that are dangerous for humans.

Robots go where humans fear to tread. Among their many applications, bomb disposal is one of the main hazardous, where the risk of death lurks with every move. Bomb disposal robots (shown in Fig. 1.2) have been used to safely disable explosive ordnance for over 40 years, where they have been deployed hundreds, if not thousands, of times. It is not just bombs that the robots dispose, but also any type of device that could detonate⑥. This could include anything from landmines to unexpected munitions.

A second example is decontaminating⑦ robots (shown in Fig. 1.3) that clean the main circulating pump housing in the nuclear power plant. Because of the radiation, humans are not allowed to do the job themselves manually, thus robots can give a helping hand.

2. Repetitive jobs that are boring, stressful, or labor-intensive⑧ for humans.

Many low-skilled jobs, which are labor-intensive as well as dull, are increasingly done by automatic machines such as robots. A good example is that welding robots (shown in Fig. 1.4) spray painting robots (shown in Fig. 1.5) do the boring jobs of welding and spray

---

④ 感知能力,这里指机器具备的感知外界和自身状态的能力。
⑤ 决策能力,指机器具备一定的自主决策能力。
⑥ 爆炸。
⑦ 净化,排除污染。
⑧ 劳动密集型。

painting instead of humans, especially in car manufacturing. Another example is pick and place robot (shown in Fig. 1.6) Remember to up-skill yourself or face the risk of lower wages and even unemployment, as new technologies like robotics and *artificial intelligence (AI)*[9] take over.

Fig. 1.2　Bomb disposal robots

Fig. 1.3　Decontaminating robots

Fig. 1.4　Welding robots

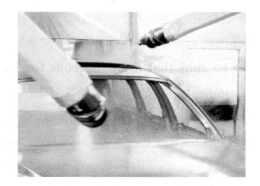

Fig. 1.5　Spray painting robots

3. Manual tasks that human do not want to do.

Asking a hundred people if they would like to do the job of toilet cleaning, chances are you will receive negative answers 100%. This is a typical work that robots can do without complaint. A typical type of toilet robot is shown in Fig. 1.7.

Thereby we can learn from the above examples that robots are there to help human beings rather than steal jobs form them. Quite the other way around, a list of brand-new, high-skilled works will be created by new technologies of robotics and automation, although being at the expense of many low-skilled works. So it is wise to learn to work with robots. AI will change everything, workers must adapt.

---

⑨　人工智能。人工智能往往结合机器人技术,比如智能机器人。

Fig. 1.6 Pick and place robot

Fig. 1.7 Toilet robot

1.3 The elements of an industrial robot[10]

Those robots essentially used in the factory production lines, hence being named *industrial robots*, find themselves ever more popular in the field of manufacturing. In *anthropomorphic*[11] terms an industrial robot requires a brain, senses, a blood supply, an arm, wrist and hand with the appropriate muscles, and possibly legs and feet, again with the associated muscles. As shown in Fig. 1.8, in a typical industrial robot, the equivalent machine elements could be a computer, measuring devices, *electric/hydraulic/pneumatic*[12] power, a *manipulator*[13] and possibly wheels.

Arms, wrists, and legs need muscles for actuation, and in practice they may be driven by pneumatic, hydraulic or electric power. Pneumatic is restricted, on the whole, to pick-and-place robots where the actuators are allowed to move quickly until arrested by mechanical end-stops. However, the compressibility of the air makes accurate control of speed and position extremely difficult, so that pneumatics are rarely found in the more demanding robot systems. For such systems the major contenders are hydraulic and electric drives. Hydraulic actuators are compact and capable of large forces and torques. They are presently popular, particularly for larger robots. Electric drives are growing in popularity with new technical advances, and becoming the most popular form of power supply today. Electricity-driven robots tend to be more accurate than their hydraulic counterparts[14], but,

---

[10] 工业机器人。有别于其他机器人，工业机器人主要用于工厂加工作业，是生产加工产品的机器人，也是本书主要的学习内容。
[11] 拟人的，将机器人比作人类。
[12] 电动/液压/气动，三种典型的机械动力源。
[13] 操纵器或机械手臂。
[14] 对应的，这里指与电驱动对应的液压驱动。

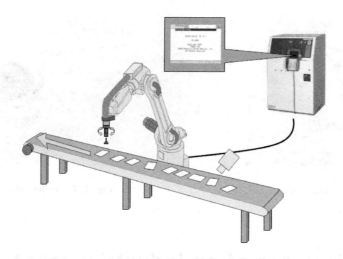

Fig. 1.8　A robot system

unlike hydraulic drives, electric motors require reduction gearboxes, thus adding to the cost of the system. A more detailed comparison of hydraulics, electrics and pneumatics will be discussed in Chapter 3.

Moving on to the sensing requirements of a robot, we distinguish two categories: internal and external sensing. If an *end-effector*[15] is demanded to move to a particular point in space, with a particular orientation, the various mechanical elements—trunk, arm, wrist—will have to be driven to the requisite positions. Measuring devices have to be installed at each *degree of freedom*[16] so that the robot knows when it has achieved those positions. This internal sensing may be carried out by *photoelectric encoder*[17] or other position/rotation measuring devices.

External sensing, on the other hand, is the mechanism of interaction with the robot's environment. First-generation robots have no such interaction, but the second-generation robots are equipped with sensors for sight and/or touch. Robot vision, based on television techniques, will allow a robot to recognize a particular component, determine its position and orientation and then command its actuators to drive the end-effector to that position. Robots with a sense of touch, possibly derived from strain gauges, will be able to react to forces generated during automatic assembly. Sensing, both internal and external, will receive further attention in Chapter 4. The brain or robot controller usually takes the form of a microprocessor or minicomputer. The controller has three main functions:

---

⑮　末端执行器,指任何一个连接在机器人边缘(关节)具有一定功能的工具。
⑯　自由度(也称为活动度 Mobility),描述某力学系统所需的独立坐标数,这里指机构能够实现的独立运动数。
⑰　光电编码器。

- to initiate and terminate motions of the manipulator in a desired sequence and at desired points.
- to store position and sequence information in memory.
- to interface with the robot's environment.

Elementary controllers, applied to sequence controls with no feedback, are often refereed to as non-servo controlled robots. In such system the initiation of each event is determined by the controller and the actual motion itself is controlled by mechanical stops which restrict the robot's motion between two end-points on each *axis*[18]. Their open-loop nature restricts them to relatively simple tasks such as transferring parts from one place to another. It should be noted that in order for such systems to qualify as industrial robots, they must be *reprogrammable*[19]. In practice, although the ordering of events is often easily done at a computer keyboard, the programmability of many non-servo controlled robots is low because of the complex arrangements of endpoints, limit switches and interlocks which determine the magnitude of the motions. On the other hand, they have several advantages such as low cost, accuracy, reliability and simplicity.

Non-servo controlled robots are not suitable for applications which require an end-effector to move to a variety of positions within a *working volume*[20]. In such cases it is necessary to use servo controlled robots. As the term implies, they are closed-loop systems in which the controlled variables need to be measured. In addition to allowing control of position, such systems can be used to control velocity and acceleration. Feedback signals are compared with predetermined values stored in the controller's memory, and the resultant error signals direct the actuators towards their targets. The end-effector can be commanded to move or to stop anywhere within its working volume. Microprocessor- or minicomputer-based programmable controllers are often used, with large memories permitting the storage and execution of more than one program, with the ability to switch to branches and *subroutines*[21]. Externally generated signals, such as those from a computer keyboard, can be used to select programs from memory.

Servo-controlled robots can be separated into two major classes—point to point (PTP) and continuous-path (CP) robots. The program for PTP control requires only the specification of the starting and finishing positions of a particular movement. In replaying the stored points the actuators on each axis, possibly six, are driven to their individual desired positions. The actual path followed by the end-effector is not easily predictable under PTP

---

[18] 机器人轴,指操作本体的轴。通用六轴工业机器人由六个伺服电机来驱动。
[19] 可重编程,即不改变硬件而能够重新编程、执行不同任务。
[20] 工作空间,指不同关节运动所达到的末端执行器的所有位置的集合。
[21] 子程序。

control unless the distance between trajectory points is made small.

There are many applications that require accurate control of the path between two points; seam welding and spray painting are typical. In such cases PTP control may be inadequate and it may be necessary to employ CP control. The CP-controlled robot may be programmed in real time by grasping its end-effector and guiding it through the required motions. By sampling[22] on a time basis, usually 60~80 Hz, data concerning position and orientation can be stored on a disc and when the data are played back the filtering properties of the robot's dynamics result in a smooth continuous motion over the desired path.

The robot's task is determined by a program which will specify the order of events and the required value of the physical variables at each event. We have already argued that the ability to change this program with ease—the programmability—is a distinguishing feature of robots. Programming can take many forms. In *manual programming*[23] the end points of each degree of freedom are fixed by physical means such as cams or limit switches, and in *teach programming*[24] all motion points are stored in memory by guiding the end-effector through the desired motions. But in addition to the two methods there is now a growing swing towards *off-line programming*[25], i.e. programming that does not require the actual robot and workpieces. Off-line programming is desirable when applications require complex motions, e.g. assembly tasks. It is also desirable in applications such as small-batch production, where it may not be possible to free a robot and make the necessary prototype parts available for teach programming. Off-line programming is also desirable when it is necessary to link the robot to other data bases, such as the output of Computer-Aided Design (CAD) systems. An off-line program requires a robot language for describing all the necessary operations in a suitable symbolic form. One of the most popular of them is VAL, a high-level language developed by Unimation to control and program their Unimate and Puma robots.

In the following Chapter 2, we will learn the different types of robots one by one.

---

[22] 取样、采样。
[23] 指最原始的手工编程方式。
[24] 示教编程。
[25] 离线编程

Chapter 2
World of robots

机器人有哪些种类？都能做些什么？各有什么特点？欢迎来到机器人大观园，让我们从最常用的生产线机器人开始。

2.1 Assembly-line robots: robots for making things

Assembly-line robots are the workhorses of robotics, i. e. , they are for making things. They are also be termed robot arms, or more technically, *multi-axis manipulators*[①]. Several examples of assembly-line robot arms are shown in Fig. 2.1, while the prototype and an model of *Selective Compliance Assembly Robot Arm* (*SCARA*)[②] are shown in Fig. 2.2 and

Fig. 2.1 Fanuc industrial robots

---

① 多轴操纵器，带有夹具等取放装置的机械手臂或机器人。
② SCARA 即选择顺应性装配机械手臂，特别适用于装配印刷电路板和电子零部件。

Fig. 2.3, respectively. Note that the first arm in Fig. 2.1 is a *parallel*[3] *Delta*[4] robot which is popular in picking and packaging in factories because they can be very fast, up to 300 picks per minute. Each robot arm consists of a base that is fixed to the ground, and a series segments each connected to the next by a joint and a motor. At the end of the final segment is a wrist joint which usually allows any one of a number of special purpose tools to be attached.

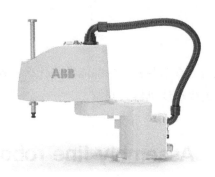

Fig. 2.2 The prototype of SCARA robot   Fig. 2.3 A SCARA robot by ABB

It is the end-effector that do the work of the robot. The joint motors allow individual *joints*[5] to be moved in different axes, which means that a combination of movements across the multiple axes will allow the end-effector to be positioned at more or less any angle or place within the reach of the robot.

Typically the end-effector can be repeatably positioned with an accuracy of better than 1 mm, which is remarkable given that the robot might weigh several hundred kilograms. Robot arms on an assembly line are typically programmed to go through a fixed sequence of moves over and over again, for instance spot-welding of car body panels, or spray-painting of car bodies. These robots are therefore not intelligent. In fact, they often have no exteroceptive sensors at all. (They will, however, normally have internal sensors for measuring the angles of each joint.) Thus a spray-painting robot may not be able to sense whether or not there is a car for it to paint. It will rely on the car, or more generally the work, to be positioned in exactly the right place, at the right time, ready for it to begin its work cycle. Thus when we see an assembly line with multiple robot arms positioned on either side along the line, we need to understand that the robots are part of an *integrated*

---

③ 并联机器人 Parallel robot 与串联机器人 Serial robot 相比,具有刚度大、精度高、承载能力强等优点,但也存在工作空间小的问题。如果比喻串联机器人为仅用一只手拿书,那么并联机器人就是两只手一起合作拿起书。

④ Delta 是一种广泛应用的高速、轻载并联机器人。

⑤ 关节,连接两连杆 Link 的连接副。

automated manufacturing system⑥, in which each robot and the line itself have to be carefully programmed in order to coordinate and choreograph the whole operation.

An important characteristic of assembly-line robots is that they require the working environment to be designed for and around them, i. e., a *structured environment*⑦. They also need that working environment to be *absolutely predictable and repeatable*⑧. Contrast this with human environments that are typically (from a robot's point of view) unstructured and unpredictable. For this reason, and the fact that assembly-line robots are dangerous to humans because they are both strong and unable to sense people, humans need to be kept at a safe distance. Although a very successful technology, assembly-line robots thus characterize the first wave of robots: robots not designed to interact with humans.

2.2 Portering robots: robots that fetch and carry

When we place an order with an online vendor, such as JD or Amazon⑨, there's a good chance that our order will be collected from its shelf in the warehouse by a robot(shown in Fig. 2.4), then brought to a central point where human packers put it in a box ready to send to us. Unlike assembly-line robots, warehouse robots are mobile. They are portering entially

Fig. 2.4　Portering robot by Amazon robotics

---

⑥　集成自动制造系统。
⑦　结构化环境,指机器人工作的周围环境场所是固定的布局。
⑧　完全可预测和可重复,指机器人作业环境必须简单、可预测;与之相对,人类所处的非结构化环境复杂而不可预测。
⑨　JD(京东)为国内最大自营电商,Amazon(亚马逊)为美国第一电商。

robots—designed ess to fetch and carry. However, warehouse robots have one thing in common with assembly-line robots: they work in highly structured environments where humans are, by and large, kept out. Some warehouses run lights-out because the robots don't need lights and humans never need to go into the main storage part of the warehouse (except, of course, if something goes wrong). Warehouse robots typically move along predefined routes through the warehouse.

However, like assembly-line robots, portering robots are highly sophisticated machines. And—also as with assembly-line robots—it is a mistake to judge a single robot in isolation. Portering robots often work in groups called *multi-robot systems*⑩ that in turn form the visible part of an integrated system of great complexity.

Warehouse or factory portering robots belong to an important class of robots called *automated guided vehicles* (AGVs)⑪. An AGV is a mobile robot that is able to autonomously navigate from one point to another, often by following a buried wire or markers on the floor. A portering AGV must be able to carry or tow⑫ a load, perhaps lifting it and setting it down automatically. A good example is a forklift AGV (in effect a driverless forklift truck): it must be able to sense obstacles in its path, including people, and safely come to a stop. An AGV is usually an electric vehicle, powered by batteries, and so it must also be able to sense its own battery level and determine when to stop working and go to a *recharging station*⑬.

We can see therefore that a portering AGV needs at least three types of sensors: to detect navigational markers, to detect obstacles, and to sense its own battery level. It also needs to communicate, via radio, with the warehouse or factory system controlling and coordinating the AGVs and, if the AGV has to be loaded or unloaded manually, it will need a simple human-robot interface so that the human operator can tell the AGV when they've finished loading or unloading. All of the requirements add up to a mobile robot that needs to be a good deal more "intelligent" than its fixed counterpart, the assembly-line robot. It also needs to be strong, reliable, and very safe.

A good example is Spot 3 robot shown in Fig. 2.5, designed by Boston Dynamics, a company specializing in animal-like robots, that is legged rather than wheeled. Spot 3 is an untethered dog-like robot that is designed to be sturdy and sure-footed when carrying supplies over uneven terrain. The project was funded by the Defense Advanced Research

---

⑩ 多机器人系统,多个机器人协同工作完成单个机器人无法完成的复杂任务。

⑪ AGV 自动导引运输车,是指装备有电磁或光学等自动导引装置,能够沿规定的导引路径行驶,具有安全保护以及各种移载功能的运输车。

⑫ 拖动,牵引。

⑬ 充电站。

Projects Agency (DARPA), and was meant to assist in dangerous situations, such as search and rescue or accessing disaster zones.

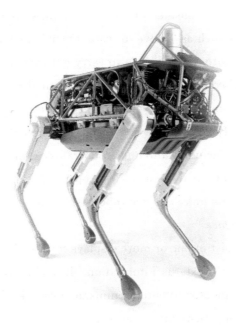

Fig. 2.5　Legged robot Spot3 by Boston Dynamics

2.3　Tele-operated robots: keeping the human in the loop

Tele-operated robots (although not necessarily named as such) have, in recent years, come to prominence in a number of important spheres of application. *Undersea remotely operated vehicles* (*ROVs*)⑭ are now a routine technology for deep-sea exploration, salvage, and maintenance. In the field of military aerospace, tele-operated *unmanned air vehicles* (*UAVs*)⑮ have since 1990 assumed considerable importance in aerial reconnaissance. Planetary exploration has had remarkable success with the tele-operated deep space probes Pioneer and Voyager, and more recently the Mars planetary rovers Spirit and Opportunity. Tele-robotic devices are finding acceptance by surgeons and are already regarded as key surgical tools for certain procedures. I shall outline examples from each of these real-world applications, but first let us define tele-operated robots.

Tele-operated robotics, or tele-robotics for short, describes the class of robotic devices that are remotely operated by human beings. Tele-operated robots thus contrast with

---

⑭　无人潜水机,能够进入深水勘探和采集样品,也称远距离遥控无人勘探机。
⑮　无人机。值得一提的是全球无人机翘楚是中国的大疆公司(DJI)。

autonomous robots that require no human intervention to complete the given task. The distinction between autonomous and tele-operated robots is blurred, however, since some tele-operated robots may have a considerable degree of local autonomy. The robot might, for example, accept commands such as "Move forward 30 cm", and automatically work out and perform the control of its motors and wheels to carry out that command, freeing the human operator from the low-level control operations. Tele-operated robots are often equipped with a rich set of sensors, including cameras, and a range of complex actuators, for both moving and manipulating objects. But when we consider tele-operated robots, it is important to consider the whole system, including the human operator, rather than just the robot.

The three main elements that define a tele-operated robot system are the operator interface, the communications link, and the robot itself. Let us consider them in turn.

(1) The *operator interface*⑯

This will generally consist of one or more displays for the video from the robot's onboard camera(s) and other sensor or status information. In addition, the interface will require input devices to allow the operator to enter commands (via a keyboard), or execute manual control of the robot (via a joystick, for instance).

(2) The *communications link*⑰

This might utilize a wired connection for fixed tele-operated robots, or wireless for mobile robots. In either case, the communications link will need to be two-way so that commands can be transferred from the operator interface to the robot, and at the same time vision, sensor, and status information can be conveyed back from the robot to the operator. Often, the communications channel will require a high capacity to carry the real-time video from the robot back to the operator.

(3) The *robot*

The robot's design will vary enormously over different applications and operating environments. But whether the robot rolls on tracks, flies, or swims, it will always need a number of common subsystems. The robot will need electronics for communication, computation, and control. It will need software to interpret commands from the operator interface and translate them into signals for motors or actuators. The software will need to monitor the robot's "vital signs", including battery levels, and send status information back to the operator alongside data from the robot's cameras and other sensors.

The defining characteristic of tele-operated systems is that the human operator is an integral part of the overall control loop. Video from the robot's onboard camera is conveyed,

---

⑯ 人机界面。
⑰ 通讯连接。

via the communications link, to the human operator. She or he interprets the scenario displayed and enters appropriate control commands that are transmitted, via the communications link, back to the robot. The robot then acts upon (or moves in) its environment in accordance with the control demands, and the outcome of these actions is reflected in the updated video data to the operator. Hence the control loop is closed.

Anyone who has tried to operate a robot remotely while peering at a screen showing the view from its camera knows how difficult it is, and good design in the operator interface is therefore very important to the success of tele-operated robot systems. A particular design problem is, for instance, how to provide the human operator with an immersive experience of the robot's sense data (i.e. as if they were there). One example of a way to improve the sense of being there is through remote *tele-haptics*[18]—in other words, allowing the human operator to feel what the robot's touch sensors are touching. A well designed tele-operated robot system acts as a kind of human sense and tool extender, to allow human exploration, inspection, or intervention in environments far too dangerous or inaccessible for humans.

From a robot design point of view, the huge advantage of tele-operated robots is that the human in the loop provides the robot's "intelligence". One of the most difficult problems in robotics—the design of the robot's artificial intelligence—is therefore solved, so it's not surprising that so many real-world robots are tele-operated. The fact that tele-operated robots alleviate the problem of AI design should not fool us into making the mistake of thinking that tele-operated robots are not sophisticated—they are. Let's now look at some representative types of tele-operated robots.

1. Autonomous underwater vehicles (AUVs)[19]

Autonomous underwater vehicles (AUVs), as shown in Fig. 2.6 and Fig. 2.7, are the workhorse robots of the offshore oil exploration and drilling industry; they are also essential robots for deep-sea exploration and science. Science AUVs famously played a key role in the exploration of the rich flora and fauna[20] around the deep-sea smoking hydrothermal vents. They are essentially remotely controlled unmanned submarines, often with robot arms equipped with grippers or manipulators so that the AUV operator can undertake maintenance or repair, or, for science AUVs, collect samples. Often AUVs are attached to the mother ship via a cable tether, which provides both power and control signals to the AUV, and the video camera feed(s) and other instrument data from the AUV back to the shipboard human operator.

---

[18] 远程触觉。
[19] 自主式无人潜水机,不需要远程操控而自主完成任务。
[20] 植物群和动物群,这里指深海的动植物群落。

Fig. 2.6 Autonomous underwater vehicle by MIT

Fig. 2.7 Jiaolong underwater robot by MST of China

2. Unmanned aerial vehicles (UAVs)

Tele-operated aircraft are generally referred to as unmanned air vehicles, or UAVs. They may be fixed wing or rotary wing, and in the military domain the best known (but controversial) UAVs are known as drones. In contrast with most other tele-operated robots, modern UAVs often have a high degree of autonomy. Fitted with a GPS receiver for satellite navigation and an *autopilot*[21], a UAV will typically be able to fly a route to a given destination, via a set of way-points, with little or no intervention from the remote pilot, much like a modern piloted commercial aircraft. UAVs will also typically be able to take off and land autonomously—something that is especially difficult to do remotely for a small aircraft. Thus a UAV operator is able to focus primarily on monitoring the vehicle's status and position, and, of course, on collecting pictures and surveillance data. It should be noted that DJI, a Chinese firm based in Shenzhen, is the leader in the field of UAV research and development. The well-known Mavic Pro UAV, shown in Fig. 2.8, was designed by DJI.

Fig. 2.8 Unmanned aerial vehicle: Mavic Pro by DJI China

3. Planetary rovers: the wheeled robots

With a distinguished history stretching back to the Lunokhod moon rover of the 1970s, tele-operated robots for surface exploration—generally referred to as *planetary rovers*[22]—have provided not only remarkable science from the surface of the moon and more recently Mars, but inspirational examples of robotics engineering. Planetary rovers are tele-operated mobile robots that present the designer and operator with a number of very difficult challenges. One challenge is power: a planetary rover needs to be energetically self-sufficient for the lifetime of its mission, and must either be launched with a power source or—as in the case of the Mars rovers—fitted with solar panels capable of recharging the rover's on-board batteries.

Another challenge is *dependability*[23]. Any mechanical fault is likely to mean the end of the rover's mission, so it needs to be designed and built to exceptional standards of reliability

---

[21] 自动驾驶。
[22] 行星漫游车。
[23] 可靠性。

and fail-safety, so that if parts of the rover should fail, the robot can still operate, albeit with reduced functionality. Extremes of temperature are also a problem, in particular the cold, since electronic components will not operate below about -50 ℃. Thus rovers, like spacecraft, need to have heaters and insulation for their electronics.

But the greatest challenge is communication. With a round-trip signal delay time of twenty minutes to Mars and back, tele-operating the rover in real time is impossible. If the rover is moving and its human operator in the command center on Earth reacts to an obstacle, it's likely to be already too late; the robot will have hit the obstacle by the time the command signal to turn reaches the rover. An obvious answer to this problem would seem to be to give the rover a degree of autonomy so that it could, for instance, plan a path to a rock or feature of interest—while avoiding obstacles—then, when it arrives at the point of interest, call home and wait. Although path-planning algorithms[24] capable of this level of autonomy have been well developed, the risk of a failure of the algorithm (and hence perhaps the whole mission) is deemed so high that in practice the rovers are manually tele-operated, at very low speed, with each manual maneuver carefully planned. When one also takes into account the fact that the Mars rovers (shown in Fig. 2.9) are contactable only for a three-hour window per Martian day, a traverse of 100 m will typically take up one day of operation at an average speed of 30 m per hour.

Fig. 2.9　Mars Opportunity and Spirit rovers

4. Surgical robots

In complete contrast to planetary rovers, the operational domain of surgical robots is

---

[24]　算法。这里指路径规划算法。

inner space. The da Vinci surgical robot shown in Fig. 2.10 is a popular case in point. Should you be unlucky enough to need prostate㉕ surgery, there's a fair chance that you will be operated on by a robot or, to be precise, a surgeon tele-operating a robot. The benefits of minimally invasive, or keyhole, surgery are well known, but the *precision and dexterity*㉖ with which the surgeon can manually control the instruments are limited. A surgical robot overcomes this limitation through the use of tele-manipulation.

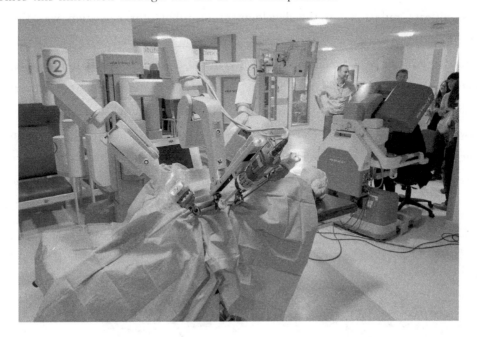

Fig. 2.10 The da Vinci surgical robot

Typically a number of laparoscopic㉗ ports are opened through the abdomen, to the site that requires surgery. Then cameras and lighting are introduced via one of the ports so that the surgeon has a close-up and possibly 3D view of the workspace. Instruments, such as miniature scalpels, forceps, or suturing needles are introduced via the other ports so that the surgeon can undertake the procedure while looking through the cameras.

The instruments are tele-operated by the surgeon via a special pair of hand controls. Importantly, the robot's control system then scales down the surgeon's hand and finger movements so that a movement of perhaps 0.5 cm by the surgeon causes a movement of 0.5 mm in the surgical instrument. The robot may also filter out any trembling in the surgeon's hands so that the micro-surgical instruments' movements are smooth and precise. Generally the surgeon's console is physically inside the operating theater, along with the

---

㉕ 前列腺。
㉖ 精度和灵巧度,指前列腺手术的极高精度和灵活性要求。
㉗ 腹腔镜的,腹腔的。

patient undergoing surgery and the whole surgical team; but, of course, it need not be: one of the benefits of this approach is that a surgeon may operate from a different location.

2.4 Robots for education

A significant milestone in the development of robots for education was the LEGO Mindstorms system, now in its second generation as LEGO NXT. Developed in collaboration with the MIT media lab, the LEGO NXT system is widely used to support the teaching of robotics and is well regarded in the robotics community. At the heart of the system is a programmable control unit called the NXT intelligent brick (shown in Fig. 2.11). As shown here, the basic Lego NXT system has, in addition to the brick, three motor units and four different sensor modules. The motor units are based on servomotors with reduction gearboxes and shaft encoders. The shaft encoders are able to sense the position of the output shaft to within 1 degree of accuracy, which means that they can be used as actuators to position the joint of a robot arm, for example, with good precision.

(a) A robot built with the NXT set (b) NXT inteligent brick

Fig. 2.11 LEGO mindstorms robot

Sensor module is a simple binary touch sensor, which outputs the value 0 if not pressed, or 1 if pressed. Sensor module is an analogue sound sensor with an output between 0 for silence and 100 for very loud sounds. Analogue light sensor module tests light level with an output between 0 and 100 (complete darkness to bright light). The module also incorporates an LED to light an object. The light sensor can be used on its own or in conjunction with the LED, to measure the brightness of reflected LED light and hence the distance to an object.

An ultrasonic[⑳] distance sensor module is able to detect the distance to an object up to a range of 233 cm, with a precision of 3 cm; the module can also sense movement. Additional

---

⑳ 超声波。

sensors not shown here are available, such as a color sensor and a temperature sensor, in addition to third-party modules including a compass and accelerometer[29].

Very simple programs can be entered directly using a simple menu system and buttons on the brick, but much more sophisticated programs can be created on a laptop or PC and downloaded to the brick. The range of programming tools (from beginner to expert) and the fact that LEGO has released an open-source licence for NXT hardware and software mean that this is a modular robotics development system of great value, not only in education, but also as a very flexible platform for testing ideas in research. An Internet search quickly reveals an extraordinary range of robot models that have been developed with LEGO NXT, including robot Rubik's cube solvers, a robot that balances on two wheels, and a fully automated LEGO NXT model factory that assembles LEGO cars.

2.5 Other Smart robots

With the development of relating technology, there are many other types of robots that are both smart and interesting.

1. AIBO toy robot

AIBO are a series of robot pets designed and manufactured by Sony. Its prototype was announced in mid-1998, and the first consumer model was introduced a year later. Although most models were dog-like, other inspirations[30] included lion-cubs and space explorer, and only the ERS-7 version and ERS-1000 versions were explicitly a "robotic dog". Fortunately, after a decade of discontinuing its support and release of AIBO robots, in 2018, Sony announced a new generation of AIBO after 11 years. This fourth generation model, ERS-1000, is expected to be smarter with embedded artificial intelligence and thus gained wider attention.

2. ASIMO robot

Honda began developing humanoid robots in the 1980s, including several prototypes that preceded ASIMO. It was the company's goal to create a walking robot. In 2000, ASIMO, the first robot to walk on two legs, was introduced. It has the ability to recognize objects, postures, gestures, its surrounding environment, sounds and faces, which enables it to interact with humans. The robot can detect the movements of multiple objects by using

---

[29] 加速度计。
[30] 灵感。

visual information captured by two camera "eyes" in its head and also determine distance and direction. Furthermore, it can not only interpret voice commands and human gestures, but also autonomously navigate thanks to sensors within its body. However, in contrast to the rebirth of AIBO(shown in Fig. 2.12 and Fig. 2.13.), ASIMO robot has been stopped development since mid-2018. But the technology behind ASIMO is expanded to use in areas such as physical therapy and self-driving vehicles.

Fig. 2.12　Sony AIBO robots

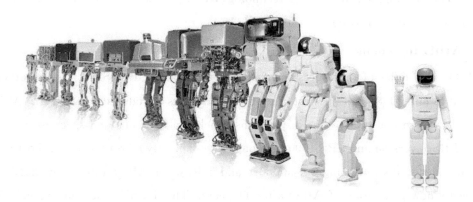

Fig. 2.13　Honda ASIMO robots

3. Sophia the robot

Although ASIMO can be treated as a humanoid robot, it still looks more like a toy that is similar to human. Sophia the robot, an advanced humanoid robot developed by Hanson Robotics, a robotics company in Hongkong, has become a global cultural icon as an almost human robot, as shown in Fig. 2.14.

Cameras within Sophia's eyes combined with computer algorithms allow her to see. She can follow faces, sustain eye contact, and recognize individuals. She is able to process speech and have conversations using a natural language subsystem. Additionally, around January 2018 Sophia was upgraded with functional legs and the ability to walk. Ultimately, she could

Fig. 2.14 Humanoid Sophia robots

be a good fit to serve in healthcare, customer service, therapy and education. Sophia runs on artificially intelligent software that is constantly being trained in the lab, so her conversations are likely to get faster, Sophia's expressions are likely to have fewer errors, and she should answer increasingly complex questions with more accuracy. Interestingly, in November 2017, Sophia was named the United Nations Development Program's first ever Innovation Champion, and is the first non-human to be given any United Nations title.

Chapter 3

Actuators for robots

任何机器都需要动力来源,机器人也不例外。与机床、生产线等大多数工厂自动化装备一样,机器人的动力主要有三个来源:气动、液压驱动、电机驱动。接下来,让我们来学习和比较这三种驱动的特点和应用。

3.1 Pneumatic

There are basically three types of power source in common use: *pneumatic*, *hydraulic* and *electric*.

Pneumatic actuators[①], as shown in Fig. 3.1, as used for industrial robots work on compressed air at a pressure of, typically, 10 bar (1 MPa), which is provided as a standard service in many factories, so the robot does not need its own compressor. They are confined almost entirely to pick and place manipulators since the compressibility of air makes it difficult to design servo systems. In a pick and place machine the valves are either fully on or fully off, and each actuator stops only at the end of its travel. A pneumatic actuator is also often used for the gripper of an electric or hydraulic robot, where its elasticity is useful as it automatically limits the force which can be applied and can cope with variations in the size of the workpieces. Also, a pneumatic gripper actuator is very light and needs to be connected only by a narrow flexible tube which is easy to feed through a complex wrist. Pneumatic auxiliary devices, particularly jigs and clamps, are often used with an industrial robot and operated by the robot controller. Another pneumatic device, although not exactly an actuator, is a cylinder which balances the force of gravity on part of a manipulator, e.g. a vertical prismatic joint. This enables a smaller motor to be used. The pneumatic compensator is just an actuator kept at constant pressure by a valve so that it exerts a constant force regardless of joint position.

---

① 气动驱动器,使用压缩空气作为动力源。

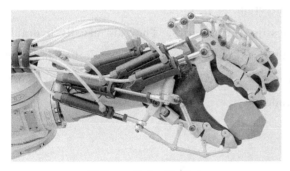
(a) Pneumatic hand by Festo

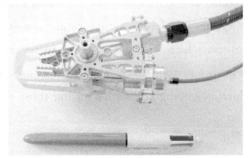
(b) Pneumatic gripper

Fig. 3.1 Pneumatic actuators for robots

The most common type of pneumatic actuator is the cylinder or ram. The other class of pneumatic actuator is the rotary actuator.

3.2 Hydraulic

As shown in Fig. 3.2, Hydraulic power[2] is used for the largest telemanipulators, some of which can carry a payload of several tons, because it is possible to generate an extremely high force in a small volume, with good rigidity and servo control of position and velocity. The high force is a consequence of the pressure at which

Fig. 3.2 Hydraulic actuator by KNR Systems

hydraulic systems are operated, 130 bar (13 MPa) being common and higher pressures not unusual. Hydraulic actuation is also commonly used for large industrial robots, although recent advances in electric motor design and gearing have made electric actuation competitive with or better than hydraulics in many cases, and most new industrial robots are electric. The basic components such as cylinders and valves are similar to those used in pneumatics, although there are many differences.

Hydraulic cylinders are usually double acting, although special forms exist, such as telescopic ones to give extra extension as in the body-tilting rams of tipper trucks. The equal area or double-rod cylinder has equal oil flows on the two sides (important in hydrostatic

---

② 液压驱动,可承受较大的载荷。

circuits) and generates equal forces in each direction.

Rotary hydraulic actuators are sometimes used in robotics. They are usually hydraulic motors, which work like a pump in reverse; several of the common pump mechanisms such as gears, pistons and rotating vanes are in use. Next let us examine electric power source which is gaining popularity during these decades.

3.3 Electric

Electric actuation[③] shown in Fig. 3.3 is the dominant method for industrial robots. Most electric robots use servomotors; these can be divided into DC servomotors, which power the majority of existing robots, and a second class sometimes called the brush less DC servomotor and sometimes the AC servomotor, which is becoming increasingly popular. Electric actuation can also be done by stepper motors, for a limited class of robots.

Fig. 3.3 Electric actuators in a KUKA robot

3.4 Other electromechanical actuators

Electric motors are, of course, electromagnetic devices. However, other transducers of electrical to mechanical energy exist, and some are used as actuators in non-robotic applications; an example is the electrostatic loudspeaker. A list of phenomena usable in actuators is given below:
- electrostatic force: electric field—force or displacement.
- piezoelectricity: deformation of a solid in an electric field.
- magnetostriction: deformation of a solid in a magnetic field.
- thermal expansion: bending of a bimetallic strip on heating.
- shape memory effect (SME): deformation of a solid on heating.

---

③ 电动驱动,是最常见的驱动方式,如电机驱动。

Others are possible, particularly if chemical reactions and fluid states of matter are considered, but the ones listed are useful because they can produce mechanical movement in a solid structure very simply. The last two rely on electric heating; it is the change in temperature which produces the movement.

These phenomena have characteristics such as small movement, low force or long time constant which make them useless for powering the main joints of an industrial robot, but there are potential applications in robotics where they can be useful. These are where a very small mechanism is needed or very precise movements must be made.

3.5 Mechanical transmission

An actuator occasionally drives a joint directly if it is small, light and powerful enough. Actuators are, however, usually heavier and bulkier than desirable, and of too low a force or torque, so *mechanical transmissions*[④] are mostly usual. The reasons for using a transmission are as follows:

- to reduce the mass, volume and moment of inertia of the machinery at the end of the arm.
- to match the speed and torque, or range of movement, of an actuator to its joint (reduction gearing and amplifying linkages).
- to use a rotary actuator to power a prismatic joint or vice versa.
- to perform mathematical or control functions—the two main examples are differential gears and parallelogram linkages.

In this section, popular power sources of robots are briefly introduced, as well as other less common electromechanical actuators that finds themselves in very specific application, e. g. Micro-Electro-Mechanical Systems (MEMS)[⑤]. With the needed power at hand, the next step is to know the robot itself and its environment, hence the important part of sensing.

---

④ 机械变速器,用齿轮及其他机械元件来变速的传动装置。
⑤ 微机电系统,是将微电子技术和机械工程融合一起的工业技术,其操控精确到微米范围。

Chapter 4

Sensing for robots

有了动力,机器人仅仅能动。它还需要一个重要的环节"检测"来感知自身和外界信息。通过检测,机器人获得自己的位置、姿态、状态和必要的外界信息。

4.1 Sensing in general

Sensing refers to the detection or measurement by a robot's controller of any physical state or quantity, from the state of a switch to a color image. It is shown in Fig. 4.1 a typical temperature sensing system of a robot hand, while those shown in Fig. 4.2 other types of sensing commonly used in a robot. Before digging into the world of robotics sensing, we introduce first sensing systems generally. Note that the terms "transducer" and "sensor" are used interchangeably. A sensing system consists of a transducer, or a combination of transducers, together with some electronics to produce a signal suitable for interfacing to the robot controller. A transducer in this context is a device which produces an electrical signal which is a measure of a physical quantity; an example is a potentiometer which produces a voltage proportional to a shaft angle. The output of a transducer is usually either a voltage or a digital signal.

Transducers are never perfectly accurate, and their specifications include several quantities expressing aspects of this. Using an angle transducer as an example, a list of specifications is obtained as follows:

- *Non-linearity*: it is the departure of the actual voltage-angle characteristic from a straight line.
- *Non-repeatability*: A possible error caused by backlash in gearing is non-repeatability.
- *Resolution*: it is the smallest interval which can be distinguished, and is often expressed as the fractional resolution.

The rest of this chapter describes the sensors most commonly used in robotics.

Sensing for robots — 4

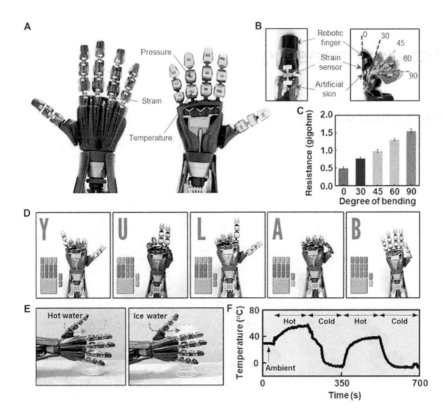

Fig. 4.1　Temperature sensing of a robot hand

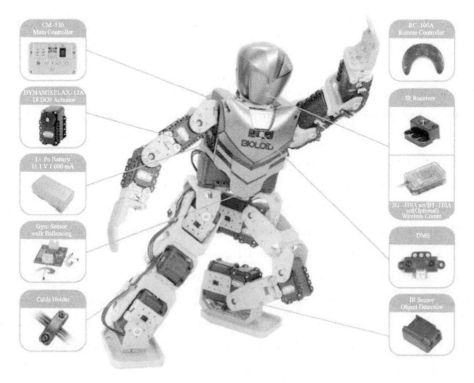

Fig. 4.2　Robotic sensors

4.2 Internal sensing

1. Joint position and rate

Joint position and rate are usually measured by an incremental or absolute encoder (shown in Fig. 4.3).

Fig. 4.3 Incremental and absolute encoders

Incremental and absolute encoders are truly digital devices. They can be made with extremely high resolution and linearity and do not depend on a precisely regulated supply voltage.

With an incremental encoder the output is a series of pulses as the encoder disc rotates. These pulses must be counted, starting from a known initial position, to give a total angle turned through. There is a danger that the count might be lost in the event of a power failure or computer restart. The alternative is the absolute encoder which has the advantage, compared to its incremental counterpart, of giving at all times a complete measurement of angle, in the form of a parallel binary signal, with no counting. But, with the additional photocells and pattern rings, and corresponding conductors and connectors, absolute encoders are more complex and expensive that incremental ones.

An incremental encoder makes a good speed sensor since the pulse rate from it is proportional to angular velocity. Speed can in principle be measured by differentiating the output of any position transducer (usually by software rather than by a circuit), but in practice this tends to give poor accuracy. Dedicated sensors for angular speed are available, in general called tachometers.

2. Force and torch

In manipulation robots, the main requirement for force or torque sensing is at the wrist, where it is used to sense forces during assembly or the force exerted on a workpiece by a tool such as a grinder held by the robot. A second place for force measurement is in the jaws of a gripper. A third category is the sensing of forces, torques and stresses at various points in the arm to determine whether safe limits are being exceeded. A common method is to use piezoelectric[①] force transducers.

A compressive force can be measured by the piezoelectric voltage generated between the opposite faces of a crystal of certain substances such as quartz[②] when it is compressed. As such crystals have little ability to support bending or tension, they must be built into a suitable structure, such as a rod sliding inside a tube, which isolates the crystal from all forces except compression, as shown in Fig. 4.4.

Fig. 4.4　A piezoelectric sensor

The piezoelectric effect also occurs in certain polymers which can be made in the form of a flexible membrane. Thus a variety of types of touch sensors can be made possible.

4.3　External sensing

1. Proximity and range sensor

Proximity sensing is the detection of the presence of an object within a certain sensitive volume. It is used in collision avoidance, detecting whether an object is near a gripper, or between its jaws, and for safety barriers; other uses are possible. Proximity detection uses sensors similar to those for short-range distance measurement; the main ones are ultrasonic and optical, often infrared.

A very simple example of object presence detection is the use of light beams between the jaws of a gripper, as shown in Fig. 4.5. Infrared optical proximity sensors can define a sensitive volume by the intersection of the field of view of the emitter and the receiver. An object within the cross-hatched area will reflect light from the transmitter to the receiver.

---

① 压电的,压电效应的。
② 石英晶体。

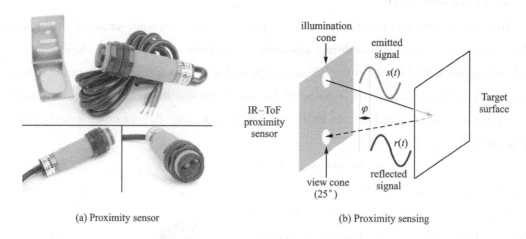

Fig. 4.5　Proximity sensor and sensing

Very dark or shiny objects may not be detected. There may be several emitters or receivers to shape the sensitive volume, or to define more than one volume. The main problem with optical sensors is interference by background lighting; this can be minimized by infrared filters and by detecting a pulsed signal from the emitter.

Ultrasonics is a good medium for object detection and range measurement up to a few meters; it is almost universal in mobile robots. Pulsed sonar at the commonly used frequency of 40 kHz has a distance resolution of about 1 cm, and a typical pulse repetition rate is 10 pulses per second. The transmitters and receivers, normally piezoelectric, are small and cheap and so can be used in large numbers.

2. Vision sensor

Robot vision is an enormous subject, particularly since it overlaps or makes use of fields such as pattern recognition and scene analysis which have many other applications. A vision system has computer-controlled cameras that allow the robot to see its environment and adjust its motion accordingly. Consequently work on remote sensing, medical imaging, military target recognition and optical character recognition may be relevant to it. Also, advances in robot vision are intimately bound up with developments in computer hardware, and much computer development is driven by the need for faster image processing. Some applications of vision in robotics are listed below:

- Detecting object presence or type.
- Determining object location and orientation before grasping.
- Feedback during grasping.
- Feedback for path control in welding and other continuous processes.
- Feedback for fitting a part during assembly.

- Reading identity codes.
- Object counting.
- Inspection, e. g. of printed circuit boards to detect incorrectly inserted components.

Vision systems are occasionally supplied by the robot manufacturer and integrated with the controller, but usually are supplied separately with an interface to the robot controller. The basic organization of a vision system is shown in Fig. 4.6. A vision sensor by COGNEX B shown in Fig. 4.7.

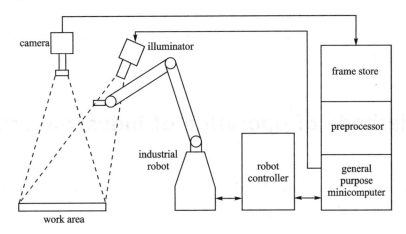

Fig. 4.6 The organization of a robot vision system

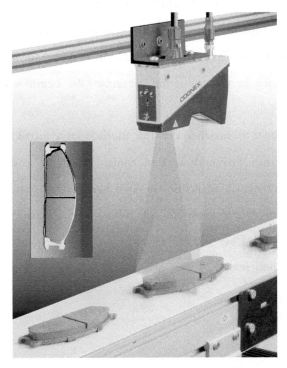

Fig. 4.7 A vision sensor by COGNEX

Chapter 5

Controlling and programming for robots

如果没有大脑,人就是一副躯干。机器人如果没有控制器,就如废铜烂铁,别指望它能干什么有意义的工作。有了控制器,还需要人通过特定的指令(编程),将要求传递给机器人。

5.1 Methods of operation of industrial robots

An industrial robot, unlike a telemanipulator, is driven through a sequence of movements by a program of some kind.

The program is executed by a controller. The controller turns on the joint actuators (throughout this chapter the terms "joint" and "axis" are used interchangeably) at the appropriate times, while signals from the joint sensors are returned to the controller and used for feedback. The types of controller, methods of programming and details of joint servo control are discussed in the following sections. We begin with the classification of industrial robots.

Industrial robots can be classified by the method of control and by the method of teaching[①] or programming; although certain control methods and teaching methods are almost always used together, in principle the two bases of classification are separate. All robot control methods involve a computer, a robot, and a variety of sensors. The main classes of control are as follows:

- Pick and place (non-servo).
- Point to point.
- Continuous path.

---

① 这里指的是机器人编程的方式。

5.2 Various coordinate systems

A *coordinate system*[2] defines a plane or space by axes from a fixed point called the origin. Robot targets and positions are located by measurements along the axes of coordinate systems. A robot uses several coordinate systems, each suitable for specific types of jogging or programming. Thus the knowledge of a set of coordinate systems and their characteristics is very important to users. There are, mainly, six coordinate systems as follows.

1. Universal or world coordinate system

The world coordinate system, always referred to as Cartesian coordinate system, is predetermined and cannot be changed. It is described with position coordinates x, y, z and rotation-angles roll w, pitch p, yaw r. All other coordinate systems are related to the world coordinate system, either directly or indirectly. It is useful for jogging, general movements and for handling stations and cells with several robots or robots moved by external axes.

2. Base coordinate system

Attached to the base of the robot, base coordinate system is only for moving the robot from one position to another. By default the base coordinate system coincides with the world coordinate system.

3. Joint or jog coordinate system

This is the coordinate system of each individual axis. Each axis rotates around its rotation center. Joint coordinate system is predetermined and cannot be changed as well. It is used for easier manual robot movement in manual mode.

4. Tool coordinate system

Tool coordinate system defines position of the tool center (TCP- tool center point). It thereby defines the position and orientation of the tool. Usually the center point of the flange in the last axis of a robot is the default TCP. The user-defined TCP is a relative distance and orientation in respect to the default TCP.

5. Work coordinate system

The system is related to the work piece and is often the best one for programming the

---

② 坐标系统。为了操控机器人,需要用到多个坐标系统,如世界坐标系、基座坐标系、工具坐标系、工件坐标系、关节坐标系、用户坐标系。

robot. It defines the placement of the work piece with respect to the world coordinate system (or any other coordinate system).

6. User coordinate system

It is defined manually by the user. It can be used for representing equipment like fixtures and workbenches. This gives an extra level in the chain of related coordinate system, which might be useful for handling equipment that hold work objects or other coordinate system.

It is easier to understand the different coordinate system with some examples.

As shown in Fig. 5.1, A is the base coordinate system for one robot, and B is the world coordinate system for reference, while C is the base coordinate system for the other robot.

Two tool coordinate systems are shown in Fig. 5.2.

A more complex example is shown in Fig. 5.3, B is the world coordinate system, C is the base coordinate system. A is the user coordinate system, while D is the moved user coordinate system, as it is defined at one corner of the table. Another user coordinate system E is defined at a point on the table; it is also called work object coordinate system. Not that if the tool or user coordinate systems are changed, the tool-path has to be reprogrammed, otherwise tool collision may occur.

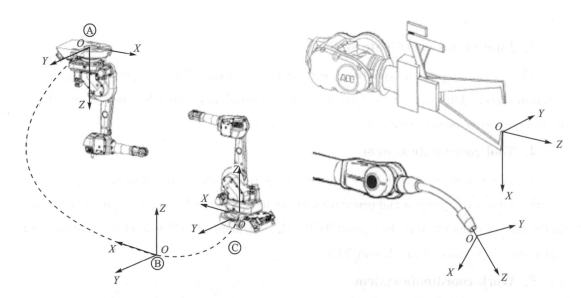

Fig. 5.1 Base coordinate systems　　　　　Fig. 5.2 Tool coordinate systems

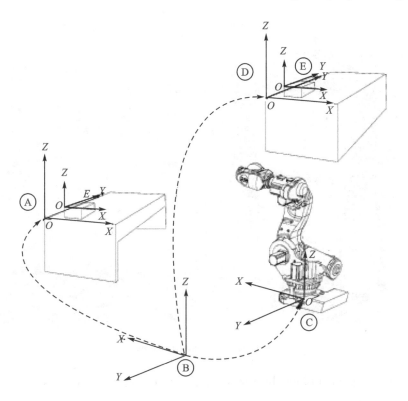

Fig. 5.3　User coordinate system w. r. t. their work and base counterparts

5.3　Methods of teaching and programming

How does an industrial robot determine what movement to make next? There are two extreme possibilities: the movement is calculated at the time, or it is replayed from an existing program or recording. The first method is necessary if the robot is to respond continuously to sensory inputs, e. g. if it is to follow a surface using a proximity sensor. Otherwise, the second method can be used. This is the method in general use. An intermediate case is for the program to have branches selected by sensor signals, or to accept certain values, such as a desired gripper rotation angle, from an external source.

1. Programming pick and place robots

The crudest form of programming is the setting up of a pick and place machine. It has two parts: the mechanical end stops are set in place for each axis, and the sequence in which the joints operate is programmed.

2. Walk-through teaching and pendant teaching

This is the most usual method with point to point servo robots. As shown in Fig. 5.4,

Fig. 5.4 Pendant teaching

a hand-held box or "pendant" has buttons, toggle switches or joysticks corresponding to each axis of the arm, which cause the axis to be driven under power (with a preset percent of velocity when the program is played back). The user drives the robot to a required position using these controls and then presses a button that causes all the joint position sensors to be read and their values stored; the robot is then driven to the next position on its required path, and so on.

3. Lead-through teaching or physical arm leading

In this form of teaching the user carries out the required motions with his own hand, while holding some device for recording the path taken, as shown in Fig. 5.5. This device may be the manipulator itself or a replica arm, the "master arm" or "teaching arm", which is geometrically similar to the robot but is light enough to move easily, is unpowered and has angular or displacement sensors on its joints similar to those on the robot. The corresponding signals are recorded and become the program which the robot plays back at a fraction or multiple of the rate compared to recorded speed.

4. Off-line programming

The alternative to teaching a robot by driving it through its cycle of operations is to type in a program at a computer terminal. In the simplest case the program consists of a series of commands of the form "move axis A through distance D". These commands are expressed in some language designed for robot programming. As shown in Fig. 5.6, Robo Guide software

Controlling and programming for robots —5

is an off-line programming tool for Fanuc robots.

Fig. 5.5　Lead-through teaching

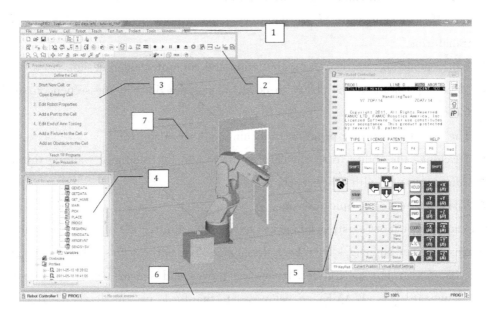

Fig. 5.6　The Fanuc programming window

5.4　Programming languages for industrial robots

Most robots can be programmed in some language which is compiled (or interpreted in the case of some slow robots intended for educational purposes) to yield the machine code which drives the robot. Many manufacturers provide a language for their own robots; meanwhile, attempts are being made to develop universal robot languages, or to add robot-control features or subroutine libraries to languages such as Pascal or C.

The main concern of this section is to explain the capabilities desirable in a robot programming language. A first requirement is that it should provide an adequate range of the facilities expected of any programming language, such as flow of control, arithmetical operations, data types, subroutines and functions. As well, there should be an adequate range of support tools (editors, compilers, debugging facilities, file handling). Also, since robot control is an instance of real-time computer control, the language and the operating system, if any, under which it runs must be suitable for real-time control; this usually implies the ability to deal with interrupts and perhaps to allow several processes to run simultaneously while communicating with each other.

In addition to these general requirements, a language for robot control should have some of the following capabilities.

- Geometric and kinematic calculations: functions and data types to allow the compact and efficient expression of coordinate systems in homogeneous coordinates and their transforms.
- World modeling: as related to simulation and computer-aided design.
- Motion specification: the ability to do the things described in the section on specifying trajectories.
- The use of sensing for program branching and servo control.
- Teaching: the ability to accept path points taught by leading or walking through.
- Communication with other machines.
- Vision and other complex sensing.

It is noted that most existing languages have only a limited set of these capabilities.

Dozens of robot languages have been developed. The earliest was MIT's language MHI in 1960. Its main robot-specific constructs are moves and sensor tests. A more general purpose language was WAVE, developed at Stanford in the early 1970s. Robot manufacturers often provide a language to go with their products. A well-known example is Unimation's VAL for the PUMA robot.

Chapter 6

Wrist, hand, and gripper

用手操控工具是人之所以区别于动物的标志。那么机器人通过什么来执行工作呢？对了，就是夹具，各种各样的夹具（记住可不是"家具"哦）。

6.1 Introduction

In anthropomorphic terms we examine the essential parts of robot. A wrist is needed to orientate the hand or end-effector in space, and the hand itself must be capable of grasping a tool or a workpiece. If mobility is desirable, the robot must be equipped with wheels, tracks or legs.

6.2 Wrist

Whilst the first three links, of a typical industrial robot of 6-axis, are used for gross position control of an end-effector, the main function of the wrist assembly is angular orientation. The mathematically ideal wrist allows rotation of the held object about three axes at right angles, such as those shown in Fig. 6.1 and Fig. 6.2. Some of the simpler robots do not have wrists: they are used in applications where tooling sets the workpiece at a fixed orientation relative to the robot. However, most tasks will require at least one angular orientation and many will require the maximum of three.

Each independent orientation of the end-effector requires a corresponding degree of freedom in the wrist assembly. There are several systems of names for the three axes; the three most commonly used terms for end-effector orientation, which is adopted here, are pitch, yaw and roll. Pitch is rotation of the end-effector about a horizontal axis at the end of the robot arm and perpendicular to its axis. It gives the end effector an up-and-down rotary motion. Yaw is rotation about a vertical axis perpendicular to the pitch axis. It gives the end-effector a side-to-side rotary motion. Finally, roll is rotation about the longitudinal axis of the wrist.

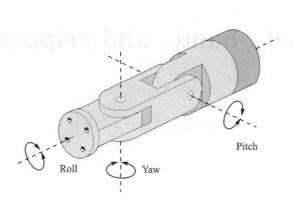

Fig. 6.1 Basic three-axis wrist

Fig. 6.2 Prototype of McGill agile wrist

6.3 Gripper

There are many ways of providing an end-effector with six degrees of freedom, but none of them serves a useful purpose until an actual payload is involved. This can take the form of a tool, such as a welding torch, a drill, a spray-gun, etc. Alternatively the payload can be an object, such as a casting, a chocolate, a sheet of glass, which is being transferred from one work station to another.

We shall bypass the rather specialized tool holder and focus our attention on the gripper, whose main function is to grasp and release workpieces in the transfer program. Typical applications are those grippers shown in Fig. 6.3~Fig. 6.6. We can classify grippers as mechanical,

Fig. 6.3 Finger gripper

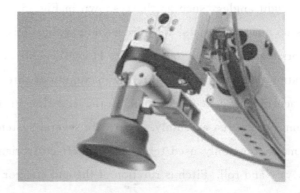

Fig. 6.4 Vacuum gripper

Fig. 6.5 Hand gripper

Fig. 6.6 Flexible gripper

vacuum and magnetic, or universal. Hence the types of mechanical gripper, vacuum and magnetic operation, universal gripper, and mechanical hands.

It is noteworthy that robot end-effectors can also be machine tools such as drills, grinding or cutting wheels, and sanders, to name just a few.

6.4 Future robotic hands

Human hands have great potentialities not only for grasping objects of various shapes and dimensions, but also for manipulating them in a dexterous manner. It is common experience that, by training, one can perform acrobatic manipulation of stick-shaped objects, manipulate a pencil by using rolling or sliding motions, perform precise operations requiring fine control of small tools or objects. It is obvious that this kind of dexterity cannot be achieved by a simple gripper capable of open/close motion only. A multifingered robot hand can therefore provide a great opportunity for achieving such a dexterous manipulation in a robotic system.

The design of multifingered robot hands has attracted the interest of the research community since the early days of robotics, not only as a challenging technical problem itself but, probably, also because of anthropomorphic motivations and the intrinsic interest for a better knowledge of the human beings. In the last decades, several important projects have been launched, and important examples of robot hands developed, as can be shown in Fig. 6.7~Fig. 6.9. Nevertheless, the current situation is that reliable, flexible, dexterous hands are still not available for real applications. For these motivations, it is easy to foresee also for the future a consistent research activity in this fascinating field, with developments at the technological and methodological level. Important connections with other scientific fields are also expected, as for example with cognitive science.

Fig. 6.7　The Utah/MIT hand

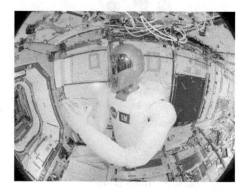

Fig. 6.8　The NASA/JPL Robonaut

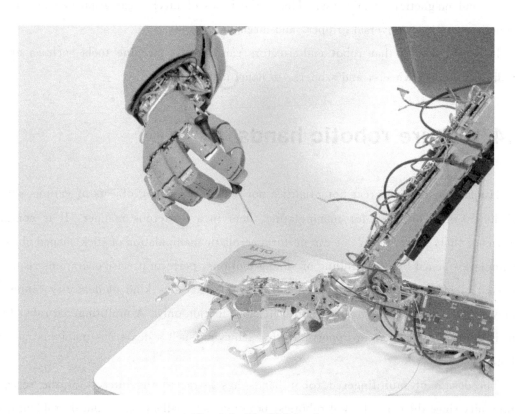

Fig. 6.9　The DLR hand

Chapter 7
Robot kinematics and dynamics

俗话说"外行看热闹,内行看门道"。看到机械手在移动,有人会说:"这个机器人好厉害!"——这个人是看热闹的;还有人会说:"这个机器人是怎么移动的,移动轨迹是什么?"这个人可能是研究运动学(Kinematics)的;而那些研究动力学(Dynamics)的则会问"是什么力让机械手如此移动? 换一种方式移动又需要什么力来驱动?"下面我们就来了解机器人的运动学和动力学。

7.1 Robot Kinematics

Kinematics[①] is the science of motion that treats the subject without regard to the forces that cause it. Within the science of kinematics, one studies the position, the velocity, the acceleration, and all higher order derivatives of the position variables (with respect to time or any other variable(s)). Hence, the study of the kinematics of manipulators refers to all the geometrical and time-based properties of the motion.

Robot kinematics applies geometry to the study of the movement of multi-degree of freedom kinematic chains that form the structure of robotic systems. The emphasis on geometry means that the links of the robot are modeled as rigid bodies and its joints are assumed to provide pure rotation or translation.

Robot kinematics studies the relationship between the dimensions and connectivity of kinematic chains and the position, velocity and acceleration of each of the links in the robotic system, in order to plan and control movement and to compute actuator forces and torques. The relationship between mass and inertia properties, motion, and the associated forces and torques is studied as part of robot dynamics.

1. Kinematic equations

A fundamental tool in robot kinematics is the kinematics equations of the kinematic

---

① 运动学。不考虑受力,研究运动的位置、速度、加速度等几何性质。

chains that form the robot. These non-linear equations are used to map the joint parameters to the configuration of the robot system. Kinematics equations are also used in biomechanics of the skeleton and computer animation of articulated characters.

Forward kinematics uses the kinematic equations of a robot to compute the position of the end-effector from specified values for the joint parameters. The reverse process that computes the joint parameters that achieve a specified position of the end-effector is known as inverse kinematics. The dimensions of the robot and its kinematics equations define the volume of space reachable by the robot, known as its workspace.

There are two broad classes of robots and associated kinematics equations: *serial manipulators*② and *parallel manipulators*③. Other types of systems with specialized kinematics equations are air, land, and submersible mobile robots, hyper-redundant, or snake, robots and humanoid robots.

(1) Forward kinematics

Forward kinematics specifies the joint parameters and computes the configuration of the chain. For serial manipulators this is achieved by direct substitution of the joint parameters into the forward kinematics equations for the serial chain. For parallel manipulators substitution of the joint parameters into the kinematics equations requires solution of a set of polynomial constraints to determine the set of possible end-effector locations.

(2) Inverse kinematics

Inverse kinematics specifies the end-effector location and computes the associated joint angles. For serial manipulators this requires solution of a set of polynomials obtained from the kinematics equations and yields multiple configurations for the chain. The case of a general 6R serial manipulator (a serial chain with six revolute joints) yields sixteen different inverse kinematics solutions, which are solutions of a sixteenth degree polynomial. For parallel manipulators, the specification of the end-effector location simplifies the kinematics equations, which yields formulas for the joint parameters.

2. Robot Jacobian

The time derivative of the kinematics equations yields the Jacobian④ of the robot, which relates the joint rates to the linear and angular velocity of the end-effector. The principle of virtual work shows that the Jacobian also provides a relationship between joint torques and the resultant force and torque applied by the end-effector. Singular configurations of the robot are identified by studying its Jacobian.

---

② 串联机械手。指机械手臂级联而成,基座与动平台之间仅有一个运动链连接。

③ 并联机械手。基座与动平台之间至少有两个独立的运动链连接。

④ 机器人雅可比矩阵,描述机器人各关节的速度与机器人动平台速度的映射关系。

(1) Velocity kinematics

The robot Jacobian results in a set of linear equations that relate the joint rates to the vector formed from the angular and linear velocity of the end-effector, known as a twist. Specifying the joint rates yields the end-effector twist directly.

The inverse velocity problem seeks the joint rates that provide a specified end-effector twist. This is solved by inverting the Jacobian matrix. It can happen that the robot is in a configuration where the Jacobian does not have an inverse. These are termed singular configurations of the robot.

(2) Static force analysis

The principle of virtual work yields a set of linear equations that relate the resultant force-torque vector, called a wrench, that acts on the end-effector to the joint torques of the robot. If the end-effector wrench is known, then a direct calculation yields the joint torques.

The inverse statics problem seeks the end-effector wrench associated with a given set of joint torques, and requires the inverse of the Jacobian matrix. As in the case of inverse velocity analysis, at singular configurations this problem cannot be solved. However, near singularities small actuator torques result in a large end-effector wrench. Thus near singularity configurations robots have large mechanical advantage.

7.2 Dynamics

We have studied static positions, static forces, and velocities; but we have never considered the forces required to cause motion. Now we consider the equations of motion for a manipulator-the way in which motion of the manipulator arises from torques applied by the actuators or from external forces applied to the manipulator.

Robot dynamics[5] is concerned with the relationship between the forces acting on a robot mechanism and the accelerations they produce. Typically, the robot mechanism is modeled as a rigid-body system, in which case robot dynamics is the application of rigid-body dynamics to robots.

There are two problems related to the dynamics of a manipulator that we wish to solve: forward dynamics and inverse dynamics. The first problem is to calculate how the mechanism will move under application of a set of joint torques. That is, given a torque vector, calculate the resulting motion of the manipulator. This is useful for simulating the manipulator. In the second problem, however, we are given a trajectory, and we wish to

---

⑤ 动力学。与运动学分析相对,动力学分析运动与力的关系。

find the required vector of joint torques. This formulation of dynamics is useful for the problem of controlling the manipulator.

Forward dynamics is also known as "direct dynamics", or sometimes simply as "dynamics". It is mainly used for simulation. Inverse dynamics has various uses, including: on-line control of robot motions and forces, trajectory design and optimization, design of robot mechanisms, and as a component in some forward-dynamics algorithms. Other problems in robot dynamics include: calculating the coefficients of the equation of motion; inertia parameter identification—estimating the inertia parameters of a robot mechanism from measurements of its dynamic behavior; hybrid dynamics—given the forces at some joints and the accelerations at others, work out the unknown forces and accelerations. Since dynamics is a more advanced topic in the robotics community, we will not elaborate on it, those interested are advised to find a textbook on dynamics or multibody dynamics for further information.

Chapter 8
Performance specification of industrial robots

百货商场商品琳琅满目,令人眼花缭乱。清楚了解你想要什么,你才能有的放矢,买到称心如意的商品。作为在工厂里从事生产作业的工业机器人,其种类繁多,如何挑选才能满足需要? 当然,首先要有一个需求及评价的标准,即性能指标。

8.1 Physical characteristics of robots

This chapter deals with the physical characteristics of robots. Physical specifications have several uses: choosing a robot for a given task; assessing whether a robot's performance has degraded over time; planning a task so that it can be done by a given robot; as targets for the design of new robots; and as a basis for designing end-effectors, including devices for enhancing the performance of the basic robots.

Because of the great variety of shapes and uses of industrial robots, standardization of specifications over all robots is difficult. However, there are certain characteristics which, all else being equal, allow robots of similar type to be compared, and these are listed in the following sections. An international standard is being prepared by the International Standards Organization (ISO). It introduces a number of conventions such as three-group of secondary minor axes (the wrist) and the end-effector. Coordinate systems are defined for the arm and the mechanical interface as well. A tentative list is as follows:

- The load-carrying capacity.
- Repeatability of positioning.
- Path tracking ability.
- Consistency of velocity.
- Positioning time.
- Static and dynamic stiffness characteristics.
- Vibrational behavior.

8.2 Geometric configuration

Robots take a bewildering variety of forms: arms of all shapes, vehicles with all possible arrangements of wheels or legs, and devices which are neither vehicles nor arms. In treating the geometric or spatial aspects of robot application, we propose that a robot is a machine for moving things around. A truly general purpose robot needs at least six controlled degrees of freedom, excluding the gripper, but, since three-axis wrists are much more difficult or expensive to make than those with two axes, cheap robots sometimes leave the third axis off. Assembly robots often use a simple pneumatic cylinder for one axis, with servo control of the others.

A machine made of rigid *links*[①] connected by joints is characterized by its number of degrees of freedom. A joint can have more than one degree of freedom. For a general industrial robot, it has six joints of one degree of freedom, hence the six degrees of freedom of the robot.

The basic kinds of motion possible at a joint are rotation and translation or sliding. A single-axis rotary joint such as a hinge is called a *revolute*[②] joint R; on the other hand, a joint with a single direction of sliding, and with no rotation, is called *prismatic*[③] joint P. Other possible joints are the *cylindrical* joint, allowing both sliding and rotation, the *helical* or *screw* joint, the *spherical* or *ball* joint. For analytical purposes these can usually be regarded as combinations of revolute and prismatic joints. Note that of all the joints found in robots, only revolute and prismatic joints are powered. Based on different combinations of joints and their configurations, robots can be classified into two types: serial and parallel. Popular six-axis industrial robots are of the serial type, as shown in Fig. 8.1; and a famous parallel one is Delta robot shown in Fig. 8.2, as well as hexapod radio telescope shown in Fig. 8.3.

Fig. 8.1 A typical serial robot

Fig. 8.2 A parallel robot: Delta

---

① 连杆,是连接两个关节 Joint 的刚性物体。
② 转动副 R,通过 R 副连接的杆件只能相对中心轴转动。
③ 移动副 P,通过 P 副连接的杆件实现某方向相对移动。

Performance specification of industrial robots — 8

Fig. 8.3 Hexapod parallel robot: a radio telescope

8.3 Positioning accuracy and repeatability

1. Accuracy

Accuracy is the ability of reaching a specified position without making a mistake. The *accuracy* with which a robot can bring the payload to a position and hold it there or the accuracy with which it passes through a position while moving, can both be important. Since this is done by servo control (expect for pick and place machines) and servos are never perfect, there will be both an offset and a random error.

Accuracy is also a function of the geometry and load at the time: the robot will tend to deflect under heavy loads and the increased inertia may affect the servos; and geometry affects accuracy, such as in the case of joint-angles. A trade-off between accuracy and speed should be taken into consideration: if more time is allowed for the servo to settle down to a commanded position, high accuracy may be obtained, at the cost of a lower overall speed.

2. Repeatability

Accuracy is a measure of how closely the robot approaches its target, on average. *Repeatability* is a measure of how closely the achieved position clusters around its means when performing a task multiple times. The difference between accuracy and repeatability is illustrated by Fig. 8.4. Repeatability is often more important than accuracy.

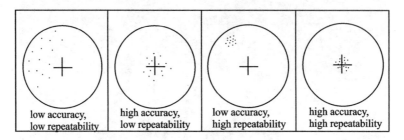

Fig. 8.4 Accuracy and repeatability

The ISO standard defines several accuracy parameters, such as
- Local pose accuracy.
- One-way pose repeatability.
- Multi-way pose repeatability.
- Stability, stabilization time.
- Overshoot on reaching a target point.
- Path-following accuracy.
- velocity fluctuation.
- Overshoot and undershoot on the transition between two straight paths.

It also distinguishes between (a) the desired *pose*[④] (b) the programmed pose, which is the robot's stored estimate of the desired pose, (c) the commanded pose, which is the control unit's interpretation of the commanded pose, and (d) the pose actually attained. Errors can arise at any stage of the chain from desired to attained pose.

8.4 Angular accuracy and repeatability

The angular accuracy of any revolute *arm* joint is the determinant of the positional accuracy for that axis. The angular accuracy of a *wrist* joint determines the accuracy with which the payload is orientated. The same apply to repeatability. Therefore, a robot specification should include the angular accuracy and repeatability of all the wrist joints.

---

④ 姿态。指机械手或参考点的位置 Position 和角度 Orientation，如末端执行器的位置和角度。

8.5 Speed and acceleration accuracy

A manufacturer's specifications will include speed, which is the amount of distance per unit time a robot can move, but this is often the maximum steady speed with the arm fully extended. In practice the arm has to accelerate and decelerate so its average speed is much lower than maximum, particularly for short strokes. The ISO standard on speed and acceleration is as follows:

- Individual axis velocity (maximum rated).
- Resultant velocity (maximum rated).
- Maximum path velocity under continuous path control, at some specified accuracy.
- Acceleration under various conditions (axis, resultant, path).
- Minimum positioning time at rated load, for a specified travel distance and path accuracy.

8.6 Spatial specifications: working volume, swept area, reach

Here we concern with the space which can be accessed by the robot. This can be expressed as the volume accessible by the payload, as shown in Fig. 8.5, and this number can be used for comparing robots, but other measurements are important as well.

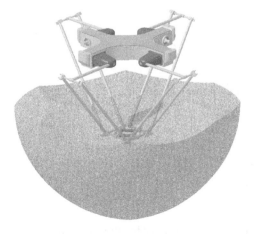

Fig. 8.5 Workspace (volume) of a robot

8.7 Payload

The payload is the maximum load capacity of a robot; it should refer to a workpiece or tool and does not include the gripper, which is regarded as part of the robot. However, with a robot with varied grippers being attached at different application, the weight of the gripper must be subtracted from the given payload capacity.

Since load affects speed and accuracy, it may be quoted for more than one condition. Load capacities range from less than a kilogram to several tons.

8.8 Vibration

Several vibrational parameters can be specified, such as
- Resonant frequencies of the robot structure.
- Amplitude of vibrations produced by the robot.
- Frequency response to applied vibration.
- Damping.
- Dynamic stiffness.
- Resistance to external vibration.

8.9 Miscellaneous specifications

Miscellaneous specifications man include
- Stiffness.
- Danger volume (swept volume including all moving parts).
- Mounting positions allowed.
- Fixing methods.
- Transport methods.
- Weight of each part.
- Cables, hoses, accessories.
- Power supplies needed (electric, hydraulic, pneumatic).

Chapter 9

Industrial robots: a case study

让我们来解剖一个典型工业机器人,看看"葫芦里面装的都是什么宝贝",各有什么作用? 工业机器人,就是面向工业领域的多关节机械手或多自由度的机器装置。

9.1 Application of industrial robots

An industrial robot is a robot system used for manufacturing. Industrial robots are automated, programmable and capable of movement on two or more axes. Typical applications of industrial robots include welding, painting, assembly, pick and place for printed circuit board, packaging and labeling, palletizing, product inspection, and testing; all these tasks are accomplished with high endurance, speed, and precision.

The biggest four industrial robot manufacturers are ABB, KUKA, Fanuc, and Yaskawa. The first two of them are established in Europe, while the remaining are both from Japan. Of them, Fanuc is indisputably the largest in the world, in terms of the number of robot arms installed in factories worldwide. It is a specialist factory automation company and is well-established in the filed of CNC as well. Thus we introduce Fanuc M-10iA/12[①] robot as a good example to learn a typical industrial robot system. As shown in Fig. 9.1, parallel robots, as opposed to serial ones, are applying into various field of industry. And dual-arm robots shown in Fig. 9.2, similar to human arms, have been designed and developed by industry leaders. A typical industry automation, as shown in Fig. 9.3, is an assembly line of car production.

---

① Fanuc Robot M-10iA Specification. https://www.fanuc.co.jp/en/product/catalog/pdf/robot/RM-10iAN(E)-02.pdf.

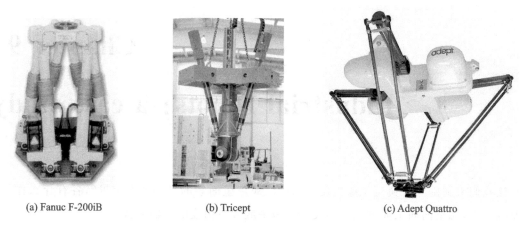

Fig. 9.1 Parallel robots into various field of industrial application

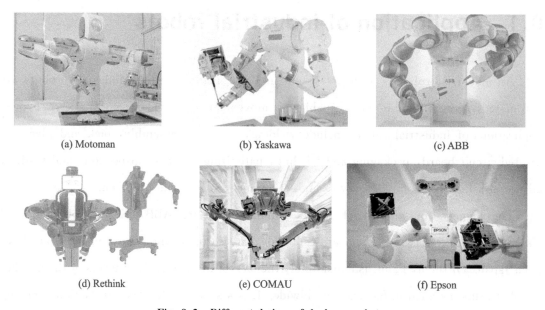

Fig. 9.2 Different designs of dual-arm robots

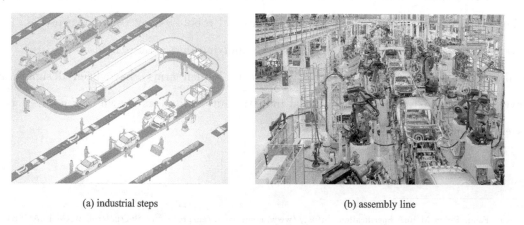

Fig. 9.3 Car production

9.2 A case study: Fanuc M-10iA/12 robot arm

1. Introduction of Fanuc M-10iA/12 robot arm

As shown in Fig. 9.4 ~ Fig. 9.6, Fanuc robot M-10iA/12 is the latest generation, six-axis, high performance industrial robot. This small but mighty robot only weights 130 kg but provides 10 kg payload with the highest wrist moments and inertia in its class. The robot comes with a motion range of 1420 mm for various application. It has the ability to position tools without vibration, even after high speed motion, achieving high speed and high accuracy at the same time and accordingly improving the productivity of the system of interest. The robot can be floor or wall mounted at any angle, or ceiling mounted, thus finding itself in various industrial solutions for: painting and dispensing, material handling and removal, assembly, picking and packing.

Fig. 9.4 Fanuc robot arm M-10iA/12

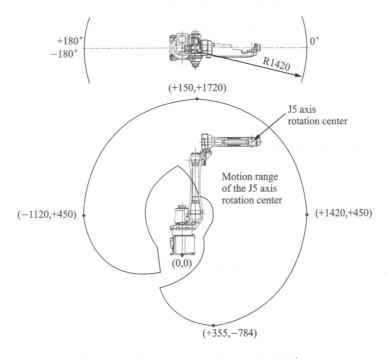

Fig. 9.5 Working range of Fanuc M-10iA/12

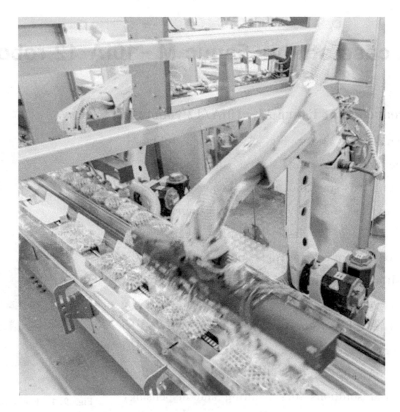

Fig. 9.6 Picking and packing workstation by M-10iA

2. Actuators, sensing, gripper, and controller

The robot are actuated by six servo-motors with integrated reducers and encoders, and can combined with iRVision sensing system for guidance and inspection.

The corresponding controller model is R-30iA, which can control up to 40 axes (robot + auxiliary). Thanks to hand cable management, off-line teaching function by Roboguide reduces teaching cost considerably.

3. Performance specification

- Controlled axes: 6.
- Robot footprint [mm]: 283×283.
- Repeatability [mm]: ±0.03.
- Mechanical weight [kg]: 130.
- Maximum load capacity at wrist [kg]: 12.
- Maximum reach [mm]: 1 420.
- Motion range [°]: J1—340; J2—250; J3—447; J4—380; J5—280; J6—540.
- Maximum speed [°/s]: J1—230; J2—225; J3—230; J4—430; J5—430; J6—630.
- Electrical connections, voltage [V]: 380~575.

- Average power consumption [kW]: 1.
- Ambient temperature [℃]: 0～45.

9.3　Picking and packing workstation

A typical application of the small but flexible M-10iA robot arm is picking and packing. Equipped with vision systems and capable of working with human-like dexterity, pick and pack robots maximize productivity in even the most demanding of automated material handling operations.

9.4　Outlook

The widespread use of industrial robots in standard, large-scale production such as the automotive industry, where robots perform repetitive tasks in well-known environments, resulted in the common opinion that industrial robotics is a solved problem. This opinion was underpinned by the robot systems' impressive automatic performance, based on advanced semiautomatic programming and resulting in an unbeatable product quality when compared to manual labor. However, large-scale production comprises only a minor part of the work needed on an industrial scale in any wealthy society, especially considering the number of companies and the variety of applications and processes that could and should be automated for productivity, health and sustainability reasons.

Global prosperity and wealth requires resource-efficient robots. The challenges today are to recognize and overcome the barriers that are currently preventing robots from being more widely used, especially in small and medium-sized manufacturing.

Taking a closer look at the scientific and technological barriers, we find the following challenges:

- Human-friendly task specification.
- Intuitive human-robot interaction.
- Efficient mobile manipulation.
- Low-cost components including low-cost actuation.
- Composition of subsystems.
- Embodiment of engineering and research results.
- Open-dependent systems.
- Sustainable manufacturing.

Chapter 10
Safety and ethics for robots

安全和伦理？没错，这是机器人专业学生需要学习的内容。任何设备都必须确保安全，工业机器人也不例外！随着人工智能的发展，机器人更新换代日益加快，哪一天机器人要是有了感情，"它"喜欢上了操控"它"的那个他或她，又该如何是好！下面就来听听机器人的安全和伦理故事。

10.1　Safety

1. Aspects of safety

Consider now of the social aspects of robotics. One of the more important of these concerns safety in the factory. Although robot installations have a much better safety record than the installations which they replace, they still present some problems. The inherent improvement in safety is a desirable feature of robots; so also is the wealth created on a global scale as a result of improved productivity. However, it would be foolish to ignore the unsettling changes in employment patterns brought about by automation.

In comparison to other machine tools, robots have a good safety record; many installations make a major contribution to safety by taking over dangerous jobs from human operators. The safety process has two stages: identifying the safety hazards and identifying systems to deal with the hazards.

2. Hazard analysis

With respect to safety, the unpredictability of robot actions is a fundamental difference between them and other machines. In 1983, Hunt sited several mistaken assumptions which operators have made about robot actions:

- If the arm is not moving, they assume it is not going to move.
- If the arm is repeating one pattern of actions, they assume it will continue to repeat that pattern.
- If the arm is moving slowly, they assume it will continue to move slowly.

- If they tell the arm to move, they assume it will move the way they want it to.

An example of the second occurs in welding applications where a robot may bring the welding tip round to a cleaning station after a fixed number of welding cycles. Again, with respect to the third, operators may not appreciate that a large hydraulic robot, normally only moving at slow speeds, is capable of violent and erratic motion simply due to a foreign particle causing a servo-valve to stick.

The *sources of hazards*[①] have been identified by Percival as:

- Control errors caused by faults within the control systems, errors in software or electrical interference.
- Unauthorized access to robot enclosures.
- Human errors when working close to a robot, particularly during programming, teaching and maintenance.
- Electrical, hydraulic and pneumatic faults.
- Mechanical hazards from parts or tools carried by the robot or by overloading, corrosion and fatigue.
- Hazards arising from the application, such as environmental hazards of dust, fumes, radiation.

There is a possibility of impact with moving parts of the robot or with items being carried by it. The greater the speed involved, the greater the danger of flying objects in the form of parts released from the gripper during turning motion, or items struck by the robot. It is clear therefore that the danger area of a robot is not confined to its working volume. Trapping points can occur within the working volume. It may be possible to be trapped between the links of the manipulator, and at all places where any part of the manipulator and its load approach other fixed items of equipment (which might even be safety guards). In this context then, the danger area is the working volume plus a clearance all round of at least 1 m.

3. Safety systems

There are three levels of protection against these safety hazards:

- Systems to prevent the operator being in a dangerous situation.
- Systems to protect the operator should this occur.
- Systems to test the second level automatically.

Guarding, usually of wire mesh or perspex, is the most common means of stopping personnel getting to the danger area but, as mentioned earlier, these structures themselves can present hazards. Guards must be well outside the working volume of the manipulator so

---

① 危险源。

that they do not form possible trapping points.

Experience shows that accidents rarely occur with robots under automatic operation: teaching and maintenance present more hazards. For these operations, access within the guard enclosure is necessary, so guard gates are interlocked with the robot control system to prevent full-speed automatic operation when a gate is opened. Multi-robot installations are guarded in two different ways. Each robot cell may be individually surrounded, with entry to a cell shutting down that particular robot.

Teaching by hand-held pendant is potentially more dangerous, since the operator usually needs to be near the end-effector to be able to position it accurately in relation to workpieces and associated machinery. Three aspects of hand-held pendants are considered to be important from a safety point of View:

- "Dead man's handle": when a pendant is released the manipulator should stop.
- The "emergency stop" button should be hardwired, and should not rely on software operation.
- Teach speeds should be restricted to safe values.

A second level of safety is necessary to take *remedial action*[2] should the first level fail or be by-passed. To detect unauthorized entry into the work area, pressure mats and/or light-beam switches are commonly used, the devices being configured to stop the robot and other machinery in the cell when triggered. In certain circumstances it may be appropriate to provide detecting devices on the manipulator itself, so that it will stop if it touches or becomes too close to other equipment. Electrical and pneumatic whiskers, infrared proximity detectors and ultrasonic devices are useful here.

It has been suggested that failsafe design may be achieved by adding a third-level system—one which tests the second level by simulating the conditions to be detected. An example has been quoted of a motor-driven vane used to break the beam to a light switch in order to test the device and its associated circuitry.

4. The safety and trustworthiness of human-robot interaction

All of the potential applications of humanoid robots, which can be broadly divided into the robot workplace assistant or the robot companion (including conversational, therapeutic, and zoomorphic robots), have one thing in common: close interaction between human and robot. The nature of that interaction will be characterized by close proximity and communication via natural human interfaces—speech, gesture, and body language.

Human and robot may or may not need to come into physical contact, but even when

---

② 补救措施。

direct contact is not required they will still need to be within each other's body space. It follows that robot safety, dependability, and trustworthiness are major issues for the robot designer. But, given that we humans are unpredictable, then how can we design and build human robots to be safe in all circumstances?

Robots, like any machine that is tasked or entrusted with a particular job, need to be designed to be safe and reliable. This is the same level of dependability we would expect from our car or washing machine, i.e. that it has been well designed, and built to meet or exceed standards of manufacture and product safety.

But making a robot safe is not the same as making it trustworthy. One person trusts another if, generally speaking, that person is reliable and does what they say they will. So if I were to provide a robot that helps to look after your grandmother and I claim that it is perfectly safe—that it has been designed to cover every risk or hazard—would you trust it? The answer is probably not.

Trust in robots, just as in humans, has to be earned. First, you would like to see the robot in action (preferably not with your grandmother). Perhaps you would like to interact with it yourself; in so doing you build a mental model of how the robot behaves and reacts and, over time, if those actions and reactions are consistent and predictable for the circumstances, then you will build a level of trust for the robot. The important thing here is that trustworthiness cannot just be designed into the robot—it has to be earned by use and by experience. Consider a robot intended to fetch drinks for an elderly person. Imagine that the person calls for a glass of water. The robot then needs to fetch the drink, which may well require the robot to find a glass and fill it with water. Those tasks require sensing, dexterity, and physical manipulation, but they are problems that can be solved with current technology.

The problem of trust arises when the robot brings the glass of water to the human. How does the robot give the glass to the human? If the robot has an arm so that it can hold out the glass in the same way a human would, how would the robot know when to let go? The robot clearly needs sensors in order to see and feel when the human has taken hold of the glass.

The physical process of a robot handing something to a person is fraught③ with difficulty. Imagine, for instance, that the robot holds out its arm with the glass but the human cannot reach the glass. How does the robot decide where and how far it would be safe to bring its arm toward the person? What if the human takes hold of the glass but then the glass slips; does the robot let it fall or should it—as a human would—renew its grip on the

---

③ be fraught with：充满…的，这里指机器人和人的交互困难重重。

glass?

At what point would the robot decide the transaction has failed: it cannot give the glass of water to the person, or they won't take it; perhaps they are asleep, or simply forgotten they wanted a glass of water, or confused. How does the robot sense that it should give up and perhaps call for assistance? These are difficult problems in robot cognition. Until they are solved, it is doubtful we could trust a robot sufficiently well to do even a seemingly simple thing like handing over a glass of water. So how might we begin to consider designing a robot that would be trusted with this kind of task?

From a technical point of view, the robot needs two control systems: one is the cognitive system that actually carries out the task. Another, parallel, safety system is one that would constantly check for unexpected faults or hazards. The primary job of the safety protection system is to stop the robot, but in a safe fashion (noting that there are some situations where freezing the robot would itself be an unsafe thing to do). First we must solve the problems of cognition and safety. Next a robot must prove itself dependable in use. Only then is it likely to earn our trust.

10.2 Robot ethics

In his 1942 short story, "Runaround", Isaac Asimov famously put forward his three laws of robotics and in so doing introduced the idea that robots could or should behave ethically. (As an aside, Asimov was also the first to coin the term "robotics".) Asimov's laws of robotics were, of course, a fictional device. Asimov himself never seriously expected that future roboticists would contemplate building them into real robots—he was well aware of how difficult this would be.

But his idea that robots should be "three laws safe" has become part of the robotics discourse. It is hard to debate robot ethics without acknowledging Asimov's contribution, and rightly so. Let us remind ourselves of his *Three Laws of Robotics*.

- First, a robot may not injure a human being or, through inaction, allow a human being to come to harm;
- second, a robot must obey any orders given to it by human beings, except where such orders would conflict with the first law;
- and third, a robot must protect its own existence as long as such protection does not conflict with the first or second law.
- a robot may not harm humanity, or, by inaction, allow humanity to come to harm.

Asimov later added *a Fourth Law*, which came to be known as the zeroth law, since logically, it precedes the first three above.

The fundamental problem with Asimov's laws of robotics, or any similar construction, is that they require the robot to make judgements. In other words, they assume that the robot is capable of some level of moral agency. To see why this is so, consider the first law: may not injure... or, through inaction, allow a human being to come to harm. "Through inaction" implies that a robot is capable of determining that a human is at risk, and able to decide if and what action is needed to prevent the possible harm. The second and third laws compound the problem by requiring a robot to make a judgement about whether obeying a human, or protecting itself, conflicts with the first (and second) law. No robot that we can currently build, or will build in the foreseeable future, is "intelligent" enough to be able to even recognize, let alone make, these kinds of choices.

Of course, even if a far-future robot were intelligent enough to make moral judgements, for it to be allowed to do so, society would have to grant it the right to be regarded as a moral agent—in other words, grant the robot personhood (or something very much like it). And with rights come responsibilities, which would again pose a difficult problem to society: what sanctions, for instance, would be appropriate for a robot that broke the robot laws? For robots to be "three laws safe" would require not only very significant advances in robot AI, but also a huge change in robots' legal status, from products to moral agents with rights and responsibilities.

Most roboticists agree that for the foreseeable future robots cannot be ethical, moral agents. (Although some, controversially, argue that near-future robots that have an artificial conscience—so that the robot's behaviors encode the military rules of engagement—could be built.) However, roboticists are also agreed that robot ethics is an important issue for the community, and several serious efforts have been made in this direction, including a European *Roboethics Roadmap*, a South Korean Robot *Ethics Charter*, and in Japan draft *Guidelines to Secure the Safe Performance of Next Generation Robots*.

So why, if robots cannot be ethical, is robot ethics an issue? There are two reasons. The first is that robots are beginning to find application in human living and working environments, and those robots will be required to interact closely with ordinary people, including children, vulnerable, or elderly people: applications that require strong safeguards regarding design, operation, and privacy.

The second is that precisely because, as we have seen, present-day "intelligent" robots are not very intelligent, there is a danger of a gap between what robot users believe those robots to be capable of and what they are actually capable of. Given humans' propensity to anthropomorphize and form emotional attachments to machines, there is clearly a danger that such vulnerabilities could be either unwittingly or deliberately exploited.

Although robots cannot be ethical, roboticists should be. Robotics researchers, designers, manufacturers, suppliers, and maintainers should be subject to a code of practice, with the founding principle that robots have the potential for very great benefit to society. What might that code of practice look like? One set of draft ethical principles for robotics proposes:

① Robots are multi-use tools. Robots should not be designed solely or primarily to kill or harm humans, except in the interests of national security.

② Humans, not robots, are responsible agents. Robots should be designed and operated as far as is practicable to comply with existing laws and fundamental rights and freedoms, including privacy.

③ Robots are products. They should be designed using processes which assure their safety and security.

④ Robots are manufactured artifacts. They should not be designed in a deceptive way to exploit vulnerable users; instead their machine nature should be transparent.

⑤ The person with legal responsibility for a robot should be attributed.

Chapter 11

Robotic futures

展望未来,机器人会发展成什么样子?

11.1 Future breakthrough on robotics

The story of robotics is still unfolding. Indeed, the pace at which robotics is advancing is accelerating, so there are likely to be significant developments in the near future, including some big surprises. By that I mean either currently overlooked corners of robotics research that turn out to be crucial, or robot technologies not yet invented that will have a huge impact.

Predicting these breakthroughs is impossible, so instead this chapter will address the question of robotic futures by outlining and discussing the technical problems that would need to be solved in order to build a number of "thought experiment" robot systems: first, an autonomous planetary robot scientist; second, a swarm[①] of medical micro-robots, and third, a humanoid robot companion with human-level intelligence.

11.2 Autonomous robot planetary scientist

Robots have a long and distinguished history in space exploration. Ever since Sputnik 1 was launched into earth orbit in 1957, space exploration has been dominated by unmanned spacecraft. Deep-space exploration has been exclusively undertaken by unmanned vehicles—think of the Voyager 1 and 2 spacecraft launched in 1977, remarkably still providing science data from the very edge of the solar system. These vehicles are in essence tele-operated robots. They are sensor-rich platforms with actuators, and a high degree of human remote control to make course corrections or to set up and operate the science instruments.

---

① 指机器人群。

Less ambiguously robotic are the planetary rovers, as shown in Fig. 11.1, which have given extraordinary service in the surface exploration of Mars, with the Mars exploration rovers Spirit and Opportunity famously far exceeding their planned mission lifetimes and science targets. There can be little doubt that robots will continue to lead the surface exploration of planetary bodies, including planetoids, moons, asteroids, and comets, perhaps paving the way for manned missions. But current planetary rovers represent the very limit of practical robot tele-operation. The inescapable physical limitation of communication delays significantly restricts the rover's speed of travel, which restricts the area that can be explored, and hence ultimately the science return from the mission. Given the very high cost of (even unmanned) planetary exploration, achieving more science for a given level of expenditure is clearly a significant driver, and one way to achieve more science is to increase the level of autonomy; in other words, build an autonomous planetary scientist.

Fig. 11.1 Planetary robot scientist

Let's begin by thinking about the perfect robotic planetary scientist. It would, ideally, be sufficiently autonomous that, once delivered safely to the surface of the planet, it would carry out its mission to explore and survey, undertaking both geology and exobiology, calling home only to send status reports, images, and science data. The robot would be capable of deciding for itself how to explore its environment and which features are worthy of closer inspection. It would be able to plan and execute whatever physical moves are needed to get itself close to those features, then decide how best to use its science instruments to find out as much as it can about them. In short, it would need to behave like a human geologist and exobiologist rolled into one, noticing and investigating anything that its human counterparts would find interesting.

But, like a lone human explorer far from home, the robot would need to exercise caution. It would need an artificial sense of self-preservation so that it can decide when not to investigate a particular object closely because the risk to itself is judged too great (think of a tall rock formation that if disturbed might collapse onto the rover, or an object in a dangerous or inaccessible place, or simply something very unexpected).

Also like a human explorer, the ideal robot explorer would need to be adaptable and resourceful. If environmental conditions change, for instance, it should be able to adapt its activity to cope, perhaps "going to sleep" to conserve energy. The rover should also be able to self-repair, so if parts of the robot become damaged, or fail through wear and tear, the robot can replace them with spares. Or if there are no spares, adapt its operation to compensate.

Of course, the science data that the rover collects is of great value, so if the robot should lose all communication and suffer catastrophic failure, it should be able to eject a data capsule, perhaps with a radio beacon, for a future rover to be able to find, rather like the explorer who perishes but leaves his diaries and notebooks to be found by those who come after.

11.3 A swarm of medical micro-robots

Our second thought experiment might appear to be pure science fiction: a swarm of microscopic robots (shown in Fig. 11.2) able to operate inside the human body and carry out medical procedures directly. The long-term vision would be a swarm that can be injected into

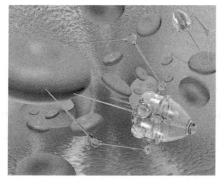
(a) micro robot

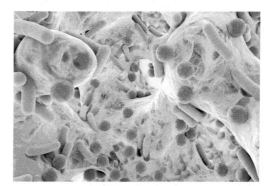
(b) microscopic robots

Fig. 11.2 Medical micro-robots

the vascular[2] system. The swarm of micro-robots would then literally swim to the source of the problem, either directed by a surgeon or with no external human control at all. Once they locate the problem, the micro-robots would ideally be able to signal their exact location and deliver therapeutics or undertake microsurgery directly. Such a technology could revolutionize medicine and, more broadly, biology and biochemistry. The ability to robotically manipulate at the level of individual cells or even molecules would open up remarkable possibilities for new science. But is this vision a fantasy, or something that might realistically be developed in the medium-term future (i.e. several decades)? And if it is feasible, what are the technical problems that need to be solved? The biggest challenge is, of course, miniaturization[3]. To get an idea of how tough this problem is, a micro-robot able to access the smallest capillaries in the vascular system would need to be no bigger than 2 micro-meters—2 millionths of a meter across. The smallest conventional robots we can fabricate are between 1 000 and 10 000 times too big. But remarkable progress has been made toward the goal of shrinking robots. Let us wait and see that progress.

11.4　A humanoid robot companion

　　For our final thought experiment, consider a humanoid robot companion. This is an important example because humanoid robots are where robots started, and in many ways a humanoid robot with human-level artificial intelligence stands as a dream of robotics. Many roboticists see such a robot as the ultimate imitation of life, and so it could be said to represent the goal of the "grand project" of robotics.

　　What do I mean by a humanoid robot companion? This robot would be an artificial companion. It would be capable of acting to some extent as a servant, a butler[4] perhaps. But rather like the butler personified by P. G. Wodehouse's Jeeves, or Andrew the robot played by Robin Williams in the movie Bicentennial Man, our butler robot must be capable of not only serving and supporting its human mistress with everyday physical tasks, as a valet, but at the same time as a provider of company: a source of entertainment, comfort, and conversation. An entity with whom one can interact, converse, or even develop a friendship.

　　Wilks, in the book *Close Engagements with Artificial Companions*, suggests that the

---

② 血管的。
③ 小型化,微型化。
④ 仆役,管家。

ideal characteristics for an artificial companion for middle-aged or elderly adults (which he labels a senior companion) might be close to the specification for an ideal Victorian companion: "politeness, discretion, knowing their place, dependence, emotions firmly under control, modesty, wit, cheerfulness, well-informed, diverting, looks are irrelevant but must be presentable, long-term relationship if possible and limited socialization between companions permitted off-duty". This implies a robot that is conversationally and linguistically highly capable, yet not overly (artificially) emotional. It must be very sensitive to its mistress's needs and feelings, demonstrate a degree of empathy but with an appropriate degree of detachment, and above all be trustworthy.

Let us set aside the question of whether such a robot companion is desirable (in the broad moral sense of whether such robots would be a good thing for society) and confine ourselves to the question of whether it is technically feasible. And assuming it is, how far is such a robot from practical realization? What technical problems would need to be solved in order to get from where we are now in terms of robot technology to where we would need to be to build such a robot?

First, consider what the robot should look like. We're assuming as a starting premise it should be humanoid, but the term humanoid covers a very wide range of physical characteristics. Should just parts of the robot, perhaps the head and its arms, be high fidelity in appearance? Or should the robot be entirely android so that to some approximation it resembles a person, like Hiroshi Ishiguro's actroid or geminoid robots?

These are not just questions of cosmetics but important considerations. The robot needs to have an appearance that is neither unsettling nor absurd, given its role as valet[5] and companion[6]: somewhere between android and cartoon-like, perhaps.

The robot's arms, and especially the hands, are likely to come into closest proximity with humans—when handing food or drink to its mistress, or assisting with dressing or washing—so very great care and attention need to be given to their appearance, softness, and compliance.

In general terms the robot needs to be small and light enough that it cannot do any real harm if it bumps into or even falls onto a human. There is some evidence that a robot should be smaller than its human mistress in order not to seem intimidating: perhaps the size of a small human adult, say about 1.5 m with a body mass of 30 kg, roughly the weight of a ten-year-old child perhaps. In addition, the robot must be soft and compliant, at least as soft and compliant as a similarly sized human.

---

⑤ 仆役,随从。
⑥ 伴侣。

Appendix A

English for Science and Technology(EST)

科技英语(English for Science and Technology)是一种学术英语和专用英语,包括科技著述、科技论文、实验报告、科技情报资料、介绍科技动向和实验的操作规程等用到的英语。科技英语要求严谨周密、概念准确,具有较强的逻辑性、客观性和严密性。因此,科技英语中通常会使用较多的长句来客观描述、准确地传递科技世界中所发生或出现的某种事情,阐释科技术语或描述某一工艺流程等。

随着科学技术的发展,机器人、人工智能、大数据、云计算、物联网等新学科不断出现,科技文献显著增多。科技英语由于其专业特性,在语法和句法上与普通英语有一些差别。这里我们将介绍使用频率较高的语法现象和词汇特点。

A.1 Characteristics of grammar and sentences of EST

A.1.1 Commonly used: passive verb

科技文献侧重叙事和推理。与一般英语相比,由于科技文献强调和突出的是作者的观点和发明,而不是作者个人的感情和价值取向;句子的重点也往往在于"谁做"而不是"做什么"和"怎么做",很多时候执行者无关紧要,于是常常使用被动语态。比如:

(1) The curve appropriate for any date is used in determining discharge for that day.

对应于每天的那条曲线可用于确定该天的流量。

(2) Every year various type of new machines are made in these factories.

这些工厂每年都制造出许多新式机器。

(3) It is demonstrated that the theory is groundless.

已经证明这一理论是站不住脚的。

(4) The production has been greatly increased.

产量已经有了很大的提高。

(5) The molecules of all matters are believed to be moving.

人们确信,所有物质的分子都在不断运动。

(6) The electric current is defined as a stream of electrons flowing through a conductor.
电流的定义是流经一个导体的电子流。

A.1.2　Commonly used: the simple present tense

科技英语常用现在时态来论述理论部分,表述不受时间限制的客观事实或普遍真理,也用此时态来说明试验的全过程。例如:

Basically, the theory proposed, among other things, that the maximum speed possible in the universe is that of light, and that mass appears to increase with speed.

基本上,这个理论,除了别的以外还提出:宇宙间能达到的最大速度是光速;质量随速度而增加。这是表示规律的宾语从句,在科技英语中表示规律、公式等的宾语从句不受主句时态的限制,仍用一般现在时。

A.1.3　Commonly used: long sentence

在科技英语中,为了明确陈述事务的内在特性和相互联系,常采用包含许多子句的复合句,或包含许多附加成分(如定语、状语、主语补足语、宾语补足语等)的简单句。科技英语中的长句通常是一个句子中有几个并列的从句或分句,这些从句或分句之间相互依附,相互制约,可以说是层层相接,环环相扣,使句子显得错综复杂,使读者眼花缭乱,不知如何处理,实难弄懂其真正意思。因此,对科技工作者来说,科技英语长句的翻译就显得越发重要。在理解和翻译这类长句时,一般用顺译法、倒译法、分译法和综合法。

1. Forward translation

Forward translation 即顺译法。当英语长句的句法结构和逻辑顺序与汉语相同或相近,且层次分明,翻译时可按原文顺序进行,一气呵成。

(1) The development of rockets has made possible the achievement of speeds of several thousand miles per hour; and what is more important it has brought within reach of these rockets heights far beyond those which can be reached by airplanes, and where there is little or no air resistance, and so it is much easier both to obtain and to maintain such speed.

火箭技术的进展已使速度可达每小时几千英里,而更为重要的是,这种进展已使火箭所能达到的高度大大超过了飞机所能达到的高度,在这样的高度上,几乎没有或根本没有空气阻力,因而很容易达到并保持火箭的那种速度。

(2) In general, drying a solid means the removal of relatively small amount of water or other liquid from the solid material to reduce the content of residual liquid to an acceptably low value.

一般来讲,干燥一种固体指的是从固体材料中除去相对少量的水或其他液体,从而使残留液体的含量减少到可接受的低值。

2. Inverse translation

Inverse translation 即倒译法。英汉两种语言在表达顺序上有着一定的差异,英语通常采用前置性陈述,先果后因,而汉语相反,通常是先因后果,层层推进,最后进行综合,点出主题。在处理某些句子时,宜采用逆序法,先译出全句的后部,再依次向前进行翻译。

(1) The construction of such a satellite is now believed to be quite realizable, its realization, being supported with all the achievements of contemporary science, which have brought into being not only materials capable of withstanding severe stresses involved and high temperatures developed, but new technological processes as well.

现代科学的一切成就不仅是提供了能够承受高温高压的材料,而且也提供了新的工艺过程。依靠现代科学的这些成就,我们相信完全可以制造出这样的人造卫星。

(2) Gas being compressed enters and leaves the cylinder through valves which a reset to be actuated when the pressure difference between the cylinder contents and outside conditions is that desired.

当汽缸内外的压力差达到期望值时,阀门开始驱动,被压缩气体才通过阀门进出气缸。

3. Separation-translation

Separation-translation 即分译法。有时原句包含多层意思,而汉语的表达习惯是一个小句表达一个意思。在翻译这类句子时,一般需要采用化整为零的分译法,就是将原文中的从句或某一短语先译出来,并通过适当的概括性词语和其他语法手段,使前后句联系在一起。整个句子可译成若干个独立的句子,其顺序基本不变,保持前后的连贯性。

(1) Steel is usually made where the iron ore is smelted, so that the modern steelworks forms a complete unity, taking in raw materials and producing all types of cast iron and steel, both for sending to other works for further treatment, and as finished products such as joists and other consumers.

通常炼铁的地方也炼钢。因此,现代炼钢厂是一个配套的整体,从运进原料到生产各种类型的铸铁与钢材;有的送往其他工厂进一步加工处理,有的就制成成品,如工字钢及其他一些成材。

(2) Manufacturing processes may be classified as unit production with small quantities being made and mass production with large number of identical parts being produced.

制造方法可分为单件生产和批量生产两类;单间生产指生产少量的机件;批量生产则是大量生产相同的零件。

4. Synthesis-translation

Synthesis-translation 即综合法。很多科技英语长句的排列顺序、逻辑关系和表达层次较汉语有着很大的差异。比如,有些长句中会有好几个并列句,而每个并列句又带有长而复杂的定语部分、主语部分或状语部分;有些长句会在句子中间部分表述事物的主要情节,而在句子的句首或句尾说出事物的背景或补充部分相关的细节。对于这种长而复杂的句子,如果只是采用顺译、倒译或分译,都会不可避免地使译文结构失调、层次混乱,从而导致理解上的错误。在这种情况下,有必要采用综合法,把原文的结构顺序全盘打乱,按其时间先后、逻辑层次和主次关系重新排列。通过这样的处理,译文会更加脉络分明,其表意更加清楚,不会造成误解。

(1) Radial bearings which carry a load acting at right angles to the shaft axis, and thrust bearings which take a load acting parallel to the direction of shaft axis-are two main bearings used in modern machines.

承受的载荷与轴心线成直角的是径向轴承,而承受的载荷与轴心线相平行的是止推轴承,它们是现代机器上使用的两种主要轴承。

(2) There is no doubt that the present drive to bring modern health services to China's peasant millions, which is daily gaining in momentum, is no temporary expedient but a long-term policy which serves the needs of today and of tomorrow.

目前,这场旨在使亿万中国农民能够享受到现代保健服务的运动正方兴未艾。毫无疑问,这并不是一种权宜之计,而是一项能满足现在和将来需要的长远的政策。

A.2　Characteristics of words

A.2.1　Specialized general words

科技文章主要是论述科技论点,或叙述某些自然规律、科学原理、现象等,而每门学科或专业都会有其特定的一套精确而含义狭窄的名词和术语。如激光(laser)、机器人(robot)。如果不懂得某一领域的专门术语,就无法理解该领域的科技文献。因此,熟悉和掌握这些科技词汇是很必要的。

很多词,在日常生活中的意思和科技英语中可能完全不同。比如,fine 在日常英语中是"好的、不错",而在科技英语中是"细小的、精细的";concrete 在日常英语中是"具体的",而在科技英语中则为"混凝土"。又比如 joint 是关节,但在科技英语中,它可能译为"接头"(supported joint)、"节理"(rock joint)、"缝隙"(expansion joint)等。

A.2.2　Prefix and suffix of words

科技词汇大多是由源于拉丁语或希腊语的词根、前缀和后缀构成的,这是科技词汇在词源方面的一大特点。了解它们将有助于我们解读和翻译英文专业书刊。据统计,在一万个普通英语词汇中,约46%的词汇源于拉丁语,7.2%源于希腊语,这些比率在科技词汇中更高。常用的前缀和后缀有100多个,并且有其独特意义。最常见的有 auto-(自,自动)、bi-(双,重)、counter-(逆,对应)、extra-(额外的)、hydro-(水)、inter-(相互,在之间)、micro-(微小的)、multi-(多)、pseudo-(伪,拟)、semi-(半,部分)、super-(超,过分)、trans-(横过,贯通)、-graph(书写物,复制的形象)、-ism(主义,学说)、-logy(学说,理论)、-scope(探测仪器)、-meter(计量仪器)、-ship(状况,性质,职业)等。用这些词缀派生、合成出的词,往往是名词,又是对事物的定义。在学习科技英语的过程中,应当尽量多地掌握前缀、后缀及其派生出的词汇,扩大词汇量,增强阅读能力,提高理解和翻译的速度和质量。

A.2.3　Abbreviation and acronym

缩略词书写方便、简洁、容易识别和记忆,在英语科技语中,有大量的缩写和缩略形式。例如,laser 是由 light amplification by stimulated emission of radiation 各词的首字母组成的;又如,ADC(Analog to Digital Converter)、3D printer(three-dimensional printer)、GPS(Global Positioning System)、QC(Quatlity Control)等。

A.2.4　Combined words

科技文章中名词和名词、名词和形容词、副词等组合的现象比比皆是,如:the intersecting through traffic lanes(相互交叉的交通车道),the normal allowable stresses(正常的容许应力值)。在通常情况下,这种组合的核心往往是最后一个名词,前面的词不论是名词还是形容词等起修饰它的作用。但有的是两个词、三个词或更多的词合在一起修饰最后一个名词。例如:a properly designed highway surface drainage system(适当设计的路面排水系统),properly 修饰 designed,highway surface 修饰 drainage,它们合起来又共同修饰 system。

与上述例子不同,还有前面的词修饰后面的词,有时也有倒置现象。例如:the electron attracting substance,其中的 attracting 是修饰它前面的 electron 的,这两个词合起来又修饰名词 substance。事实上我们还可以把这两个作定语的修饰词,转换成定语从句来修饰 substance,如:the substance which attracts the electron(吸引电子的物质)。

其他如 flow-sheet(流程图,名词+名词)、stereo image(立体图像,形容词+名词)、insulating material(保温材料,分词+名词)这类合成词的理解和翻译多采用直译法,要进行语义分析、准确找出中心词、正确理解组合结构,再准确表达原意就水到渠成了。

Appendix B
How to plan your career and create a resume

无论是对刚毕业的学生还是对久经"沙场"的专业人士,好的简历是找到好工作的第一步。而职业规划将影响整个职业生涯的发展,它包括职业定位、目标设定和通道设计三要素。职业规划不能按部就班,应该从自身客观环境分析开始,量身定制。做好了规划,接下来就是求职。求职始自简历制作。一份简历在招聘经理眼里只停留几秒,好的结构和有序组织的内容能让你的简历脱颖而出。想要成为求职大军上强有力的竞争者,需要做到两点:①根据岗位准备个性化简历;②抓大放小,突出核心技能、教育背景和工作背景。接下来,我们来一步步了解如何制订适合自己的职业规划,并制作一份优秀的英文简历。

B.1 How to plan your career?

A career plan lists short- and long-term career goals and the actions you can take to achieve them. Career plans can help you make decisions about what classes to take, and identify the extracurricular activities, research, and internships that will make you a strong job candidate. Below are some helpful steps to guide you in creating a career plan customized to your interests and ambitions.

1. Identify your career options

Develop a refined list of career options by examining your interests, skills, and values through self-assessment. Narrow your career options by reviewing career information, researching companies, and talking to professionals in the robotics field. You can further narrow your list when you take part in experiences such as shadowing, volunteering, and internships.

2. Prioritize

It is not enough to list options. You have to prioritize. What are your top skills? What interests you the most? What is most important to you? Whether it is intellectually challenging work, family-friendly benefits, the right location or a big paycheck, it helps to know what matters to you – and what is a deal-breaker.

3. Make comparisons

Compare your most promising career options against your list of prioritized skills, interests and values.

4. Consider Other Factors

You should consider factors beyond personal preferences. What is the current demand for this field? If the demand is low or entry is difficult, are you comfortable with risk? What qualifications are required to enter the field? Will it require additional education or training? How will selecting this option affect you and others in your life? Gather advice from friends, colleagues, and family members. Consider potential outcomes and barriers for each of your final options.

5. Make a choice

Choose the career paths that are best for you. How many paths you choose depends upon your situation and comfort level. If you are early in your planning, then identifying multiple options may be best. You may want several paths to increase the number of potential opportunities. Conversely, narrowing to one or two options may better focus your job search or graduate school applications.

6. Set "SMART" goals

Now that you have identified your career options, develop an action plan to implement this decision. Identify specific, time-bound goals and steps to accomplish your plan. Set short-term goals (to be achieved in one year or less) and long-term goals (to be achieved in one to five years).

- Specific—Identify your goal clearly and specifically.
- Measurable—Include clear criteria to determine progress and accomplishment.
- Attainable—The goal should have a 50 percent or greater chance of success.
- Relevant—The goal is important and relevant to you.
- Time bound —Commit to a specific timeframe.

7. Create your career action plan

It is important to be realistic about expectations and time-lines. Write down specific action steps to take to achieve your goals and help yourself stay organized. Check them off as you complete them, but feel free to amend your career action plan as needed. Your goals and priorities may change, and that is perfectly okay.

8. Meet with a career advisor

Each school has career advisors who are there to help you make effective career decisions. Make an appointment on campus to talk about your career options and concerns.

B.2 How to create a resume?

B.2.1 Structuring your resume

1. Choose a template or make it yourself

Most word processing apps have several different resume templates that you can choose from. If none of them appeal to you, however, you can always use your own design. There are also templates available for download online, mostly free. If you do not want to use one of the basic templates in your word processing app, you may find another online that works for you. Template elements can also be customized to suit your needs. Think of it as scaffolding that you can adjust or eliminate as necessary. Remember:

- Use a standard, readable font in 10- or 12-point. Your section headings may be a little larger.
- Times New Roman and Georgia are popular serif fonts. If you want to go with a sans-serif font, try Calibri or Helvetica.

2. Header with your name and contact information

At the top of your page, type your full name, address, phone number, and email address. Play around with the formatting to find something that you like best. For example, you could have all the information centered. You could also have your address on the left side and your phone number and email address on the right, with your name centered in the middle in a slightly larger size. If you don't already have a professional email address, get one from a free email service such as NetEase. Ideally, the email address you use on your resume will be some version of your initials and last name. Never list a silly or suggestive personal email address on your resume.

3. Conservative fields: chronological resume

In a chronological resume, you list your work experience and education in reverse-chronological order. This is a classic resume format that would likely be more appreciated by older hiring managers, or those in conservative fields such as accounting or law. You don't have a lot of flexibility with a chronological resume, but you can still arrange the sections in a way that puts your strongest information at the top. For example, if you have a lot of education but not a lot of work experience, you might want to list education first.

4. Lack direct work experience: functional resume

With a functional resume, you can highlight your specific skills and assets without having to list every job you have ever had. This can be a benefit if your work experience is thin. A functional resume is also a good choice if you have an extensive amount of experience and want to limit your resume to a page. You can focus on the skills you have developed rather than having to list each individual job with specific details.

5. Highlight skills: combining chronological and functional resume

You can still use a functional resume even if you are looking for a job in a more conservative field. Lead off with a skills section, then include chronological sections underneath. Because this type of resume can get lengthy, consider only listing your last jobs and your highest educational degree. You can make clear in the functional part of your resume how long you have been working in the industry.

B.2.2 Making your content shine

1. Functional resume

A functional resume puts emphasis on what you can do, rather than on what you have done. Brainstorm a list of 4 to 5 categories of skills that you have experience or education in. Then include a brief description of the skill and bullet points with specific examples of when you put that skill to use.

You can include relatively soft skills. For example, you might list "team leader" as one of your skills. You could then include bullet points detailing your work in student government, organizing a rally for a nonprofit organization, or working as a camp counselor.

2. List all relevant working experience, including volunteer work

For a chronological resume, add specific jobs and other work experience in reverse-chronological order, starting with the most recent job you have had. Use specific, descriptive job titles that tell potential employers exactly what you did through that experience. Use active verbs to describe your responsibilities and accomplishments. Specific numbers and metrics show potential employers exactly what you achieved.

3. Adding education and certifications

Typically, your highest degree is the only one you need to include on your resume. However, you should include lower degrees if they are relevant to the job you are applying for. Additionally, list any relevant licenses or certifications you have. If you had a cumulative GPA of 3.5 or higher, feel free to include it with your educational information. Otherwise, you should leave it out. If you are listing more than one degree, include your GPA on both of them if possible. Otherwise, do not include it at all.

4. Emphasize hard skills

Focus on hard skills that can be objectively evaluated, such as computer or technical skills or languages. If specific skills were listed in the job post and you happen to have those skills, on the other hand, include a skills section and list them along with detail about your proficiency.

5. Keywords

Employers often use filtering software to scan resumes for specific keywords. These words indicate what they want in a potential employee. The software enables hiring managers to spend less time reading through resumes. To get through the filter, include keywords mentioned in the job listing. But make sure the keywords you use fit in with the rest of your text, and use them sparingly.

6. Hobbies and interests

Here the rule is that you should include only hobbies or interests that would benefit you in the job you are applying for.

B.2.3 Finalizing your resume

1. Tailor-made your resume

You may have a master resume that includes all of your skills, education, and experience. However, the resume you give to each potential employer will not necessarily include everything. Only include skills and experience that are directly related to that job. Try to make your resume match the job listing as closely as possible.

2. Save space and avoid additional words

Active, punchy text is essential in a resume. Remove pronouns, articles, adjectives, and adverbs. The final statement should communicate only the action and the result of that action.

3. Double-check before submitting

Do not rely solely on the grammar and spelling checkers built into your word processing app. Read through your resume several times to make sure it is indeed error free. Reading out loud can also help you find errors or awkward wording.

4. Save as a PDF file

Use PDF file format unless the job listing specifically requests something else.

5. Print copies of your resume

Print your resume on a good printer using quality white or ivory paper. You can find "resume paper" online or at office supply stores. If you included hyperlinks for your digital resume, take them out before printing so all your text will be black.

Take at least three copies of your resume to the interview. If you know you are being interviewed by a hiring team, take enough copies so that each member of the team can have one. You also want to make sure you have at least one left over for yourself.

B.2.4 Resume template

Next, please share the following three resumes downloaded from the WikiHow website, they are functional, chronological, and combinational, respectively.

Introduction to Robots and Robotics ◎ 机器人专业英语

Template 1

Christina M. Shaffer
212 South Jefferson Street
Georgetown, TX 78626
Phone: 512.686.2398
Email: cmshaffer@email.com

EDUCATION

Texas State University, San Marcos, TX   1994—1998
Degree: Bachelor of Political Science
Grade point average 3.97

LEGAL EXPERIENCE

Paralegal   May 2009—present
Jenkins Law Office, Georgetown, Texas

Schedule appointments. Draft legal documents including but not limited to estate planning, probate, guardianships, guardianship accountings, quitclaim, personal representatives, and warranty deeds. Prepare inheritance tax returns and consents to transfer, file pleadings, and record deeds. Correspond with clients and courts. Organize and maintain client case files.

Paralegal/Office Manager   January 2000—May 2009
Howe, Logan & Marx, San Marcos, Texas

Scheduled appointments. Drafted legal documents including but not limited to dissolution, custody, child support, adoption, guardianship, motions, petitions, and orders. Corresponded with clients, courts, and other attorneys. Scheduled hearings. Organized and maintained client case files. Conducted legal research. Implemented and managed office website and software. Maintained accounts receivable and payable.

BUSINESS EXPERIENCE

Self-Employed   May 1995—January 2000
Georgetown, Texas

Designed, managed, and advertised website and online businesses for various clients. Duties included ASP/CGI/Java installation and modification, sales management, scheduling client's business meetings, corresponding with client's customers, assessing advertising needs, billing, graphic design, and information and market research.

Template 2

Sarah R. Holding
619 Main Street | Butte, MT 59703 | (406) 494-6221

EMPLOYMENT HISTORY

Executive Assistant to Vice President　　　　　　　　　　　　August 2009—Present
XYZ Corporation, Butte, MT

Serve as liaison between departments and operating units in the resolution of day-to-day administrative and operational problems.

Major Accomplishments
- Made international travel arrangements for senior-level executives via the Internet, resulting in average net saving of $250 per person/per trip.
- Dispatched messengers on assignments, coordinating trips to ensure that multiple stops were made each time. Saved the company approximately $49.75 per messenger per day.

Senior Administrative Assistant　　　　　　　　　　　　November 2007—August 2009
ABC Magazine, Butte, MT

Composed and edited correspondence and memoranda from dictation, verbal direction, and knowledge of departmental policies. Prepared, transcribed, and distributed agendas and minutes of numerous meetings.

Major Accomplishment
- Created reliable and efficient client database, saving the company approximately $4,500 in technical support expenses.

Secretary/Administrative Assistant　　　　　　　　　　　　July 2002—October 2007
Raymond Pool Systems, Butte, MT

Scheduled and coordinated appointments and events for supervisors. Entered client information and financial data into computer system.

Major Accomplishment
- Implemented client data and file management system, saving the company $65,000 in the first year of use.

EDUCATION HISTORY

Hayes Business College, Butte, MT　　　　　　　　　　　　2000—2002
Associate of Science in Administrative Assisting

Downloaded from http://www.wikihow.com

Template 3

SALLY REYNOLDS
227 E. 4th Street
Columbus, OH 43210
Email: sreynolds@email.com
Phone: (614) 292-2350

EDUCATION

Bachelor of Science in Sociology
Ohio State University, Columbus, OH
GPA: 3.85

RESEARCH INDUSTRY EXPERIENCE

Research Assistant
Taylor University, London, OH
- Prepared tables, graphs, fact sheets, and written reports summarizing research results.
- Prepared, manipulated, and managed extensive databases.
- Provided assistance in the design of surveys.

FINANCIAL INDUSTRY EXPERIENCE

Financial Assistant
Reinhold & Reinhold, LLP, Columbus, OH
- Entered new customer data into customer/accounting system.
- Corresponded with clients, the Internal Revenue Service, and the Ohio Department of Revenue.
- Faxed information as required.

CUSTOMER SERVICE EXPERIENCE

Receptionist
Schmidt Accounting, Columbus, OH
- Greeted clients, organized client files, handled multi-line phone system, and scheduled appointments.

AWARDS & RECOGNITIONS

American Association of Research Professionals
Outstanding Researcher of the Year 2009

Downloaded from http://www.wikihow.com

Appendix C

Code of ethics for robotics engineers

机器人产业飞速发展,从业人员需求急剧增加,作为即将从事机器人相关工作的"准员工",需要具备什么样的职业素养和职业道德呢?很明显,结合了人工智能的机器人具有"类人"特征,它能与环境互动并作出判断,进而采取合适的行动。这就要求设计、开发、应用、维护机器人的上下游从业人员具备高于传统行业人员的道德和素养。参照国际电气和电子工程师协会(IEEE)职业道德规范,Worcestoer Polytechnic Institute(WPI)的 Brandon Ingram 等提出了机器人行业道德准则,这里将全文摘出供同学们参考。

C.1 Preamble

As an ethical robotics engineer, I understand that I have the responsibility to keep in mind at all times the well-being of the following communities:

- Global—the good of people and known environmental concerns.
- National—the good of the people and government of my nation and its allies.
- Local—the good of the people and environment of affected communities.
- Robotics Engineers—the reputation of the profession and colleagues.
- Customers and End-Users—the expectations of the customer and end-user.
- Employers—the financial and reputational well-being of the company.

C.2 Principles

To this end and to the best of my ability, I will ...

1. Act in such a manner that I would be willing to accept responsibility for the actions and uses of anything in which I have a part in creating.

It is the responsibility of a robotics engineer to consider the possible unethical uses of the engineer's creations to the extent that it is practical and to limit the possibilities of unethical use. An ethical robotics engineer cannot prevent all potential hazards and undesired uses of the engineer's creations, but should do as much as possible to minimize them. This may include adding safety features, making others aware of a danger, or refusing dangerous projects altogether. A robotics engineer must also consider the consequences of a creation's interaction with its environment. Concerns about potential hazards or unethical behaviors of a creation must be disclosed, whether or not the robotics engineer is directly involved. If unethical use of a creation becomes apparent after it is released, a robotics engineer should do all that is feasible to fix it.

2. Consider and respect peoples' physical well-being and rights.

A robotics engineer must preserve human well-being while also respecting human rights. The United Nations' Universal Declaration of Human Rights (http://www.un.org/en/documents/udhr/index.shtml) outlines the most fundamental of these rights. Privacy rights are especially of concern to a robotics engineer. A robotics engineer should ensure that private information is kept secure and only used appropriately. There are circumstances when honoring privacy rights or other rights conflict with preserving the well-being of an individual or group. In these cases, a robotics engineer must decide the ethical course of action, making sure the least harm is done.

3. Not knowingly misinform, and if misinformation is spread do my best to correct it.

A robotics engineer must always remain trustworthy by not misinforming customers, employers, colleagues or the public in any way. A robotics engineer must disclose when the engineer feels unqualified to safely or fully complete a required task. When others spread misinformation, a robotics engineer must do as much as possible to correct the misinformation.

4. Respect and follow local, national and international laws wherever applicable.

A robotics engineer must follow the laws of the applicable communities. This includes where the robotics engineer is working and the communities targeted by the outcome of the engineer's work. The intellectual rights of others should be maintained at all times and assistance received from others should always be properly credited.

5. Recognize and disclose any conflicts of interest.

A robotics engineer must disclose the existence of any conflicts of interest to employers. It is up to the robotics engineer to decide how to react to any such conflict, either by attempting to ignore personal feelings or by avoiding the source of conflict.

An employer must be aware of conflicts and that these conflicts of interest may affect the robotics engineer's decisions. Bribery inherently creates conflicts of interest and is unethical.

6. Accept and offer constructive criticism.

A robotics engineer should always strive to produce the best work possible and to help others do the same. For this reason, a robotics engineer must both give and accept constructive criticism. This allows for robotics engineers to help improve each other's work, benefiting each other and those affected by the robotics engineer's work. A robotics engineer who refuses to consider criticism risks making avoidable mistakes.

7. Help and assist colleagues in their professional development and in following this code.

This code of ethics is available as a guideline for all robotics engineers as a means of uniting them with a common basis for ethical behavior. In following this code, a robotics engineer promotes the positive perception of the field by customers and the general public. In helping colleagues develop professionally and ethically, a robotics engineer makes sure that the field of robotics will continue to grow.

C.3 Conclusion

This code was written to address the current state of robotics engineering and cannot be expected to account for all possible future developments in such a rapidly developing field. It will be necessary to review and revise this code as situations not anticipated by this code need to be addressed.

Appendix D
Information retrieval

在学校里，常常会有老师布置一些课程项目和课题任务让大家去做？不少同学接到任务和作业后，开始急匆匆，中间累得慌，最后草草收尾。走了不少弯路，为伊消得人憔悴，蓦然回首，才发现那人却在灯火阑珊处。做到最后才发现，原来早有现成的方案可供参考，忙痛根自己为啥不知道还有文献检索这个乾坤大法！

互联网的飞速发展，让信息的实时共享变成了现实。小到做一个课程项目，大到开发大型系统或完成毕业设计之前，要了解别人是如何做的，借鉴前人的研究和开发经验，避免走弯路的办法就是先检索文献。

D.1 Information retrieval，more important than you thought

D.1.1 What is information retrieval

什么是文献检索呢？百度百科这么来定义（见图 D.1）：文献检索（Information Retrieval or Literature Retrieval）是指根据学习和工作的需要获取文献的过程。近代认为文献是指具有历史价值的文章和图书或与某一学科有关的重要图书资料。随着现代网络技术的发展，文献检索更多是通过计算机技术来完成。

要想知道更多关于文献检索的定义，请大家打开任何一个网页浏览器（IE 或 Chrome 或 FireFox 等），键入百度百科网址"https://baike.baidu.com"，在搜索框内输入"文献检索"词条，单击回车键之后，答案就呈现在面前。这个过程就是文献检索。接下来我们来做一个课堂作业，开始第一次文献搜索。回到刚才的百度百科页面，在搜索框内输入"工业机器人"，等待若干秒，是不是看到了图 D.2？恭喜你，你成功完成了一次文献检索。

D.1.2 What can information retrieval do？

当代科技进步飞速，科学与技术的发展又具有连续性和继承性，关起门来搞研发只会重复别人的劳动、走弯路。研究人员在选题开始就必须进行文献检索，了解前人在该项目上已经做的工作、目前正在做的工作、由谁在做、在怎么做。

图 D.1　文件检索在百度百科的定义

图 D.2　检索"工业机器人"

大家要养成一个习惯:碰到新问题,首先检索文献,了解背景知识。文献检索是研究工作的基础环节,掌握了文献检索的方法,就等于找到了一条吸收和利用大量新知识的捷径。一般来说,文献检索有以下三大作用:

① 避免重复研究;

② 节省时间;

③ 获取新知识。

D.1.3 How to do the literature searching?

进行检索前,首先要知道有哪些(种类)文献。按照文献的出版形式,文献大致可以分为以下几种类型:

- 图书(品种最多、数量最大、范围最广);
- 期刊(周期短、报道快、数量大、内容新);
- 报纸(报道及时、受众面广、通俗易懂);
- 学位论文(较为系统,有深度有创造);
- 会议论文(专业性强、方向前沿);
- 专利文献(新颖、实用、详尽);
- 标准文献(各种规格、规则、技术要求,如国标);
- 科技报告(单独成册、内容专深、报道迅速、控制发行);
- 政府出版物(官方或半官方机构发表的行政性、科技性文件);
- 产品资料(商业性宣传资料);
- 其他文献(包括技术档案、广播、电视、简报、复印资料等)。

那么如何检索文献呢?通常检索文献分为以下 6 个步骤:

① 知道课题或项目要求;

② 了解课题或项目的知识背景;

③ 分析课题或项目涉及的知识和概念;

④ 选择合适的文献类型(如适用的刊物或数据库);

⑤ 实施检索;

⑥ 获取文献原文。

下面分别通过"CNKI"和"ISI Web of Knowledge"来说明中文数据库和英文数据库的检索和使用。

D.2 Retrieval of Chinese database:CNKI

中国知识资源总库(CNKI)简称中国知网,是目前最常用的中文数据库。下面我们以它为例来学习如何使用中文数据库检索信息。

要进入中国知网首页,只要在网址栏键入"https://www.cnki.net/",在国内任何地方,登录该网站即可进行检索,但大多数情况下无法阅读全文。如果你所在的城市(如杭州)订阅了中国知网,那么在杭州市范围(IP 地址控制),就可以下载全文。当然,基本所有国内的大学和专科院校都订购了中国知网(单位订阅),因此你可以在校园内检索和访问 CNKI 资源,并下载全文。

接下来,我们来做一次搜索。

① 打开知网首页，如图 D.3 所示。

图 D.3 中国知网 CNKI 首页

左侧的检索框用来选择文献的来源，有全文、主题、篇名、作者、单位、关键词、摘要、参考文献、中图分类号、文献来源等，默认方式是"主题"。

② 选择默认方式。

③ 在右侧搜索框内输入"工业机器人"。

④ 单击搜索框右侧的"搜索"键，出现如图 D.4 所示页面，显示按照发表时间排序（默认排序）的文献检索结果。

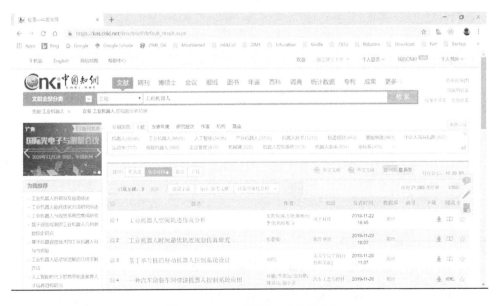

图 D.4 CNKI 检索"工业机器人"

按发表时间的排序,把最新发表的文献放在最前面,以便于读者了解该领域最新的研究动向。有时候我们想要全面了解某个领域的研究情况,了解前人在该领域所做的主要工作,领域内主要的研究机构和研究人员,我们可选择:

⑤ 按"相关度"排序结果,把"工业机器人"领域最相关的文献优先排位,如图 D.5 所示。

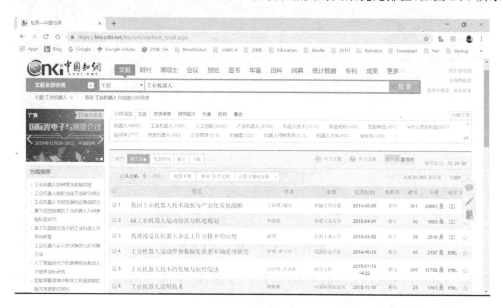

图 D.5 "工业机器人"检索结果按相关度排序

⑥ 我们也可以选择"被引"这个选项,按被引用次数来排序文献,如图 D.6 所示。

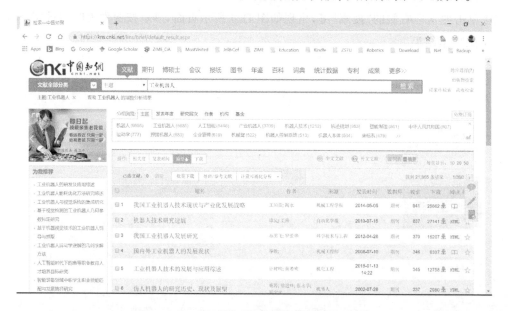

图 D.6 "工业机器人"检索结果按被引次数排序

被引次数多的文献,一般是领域内大牛及所在团队的成果,行业内认同度最高,通过被引次数排序,可以快速定位所选领域的权威作者和单位。

当然，我们也可以选择以下方式：

⑦ 按下载次数排序（如图 D.7 所示），得到的结果和被引次数差别不大。

因为里程碑式论文不仅被引次数高，其被下载的机会也最高。通过优先下载和学习最核心、里程碑式的文献，可以大大缩短我们检索的时间，提高检索效率。

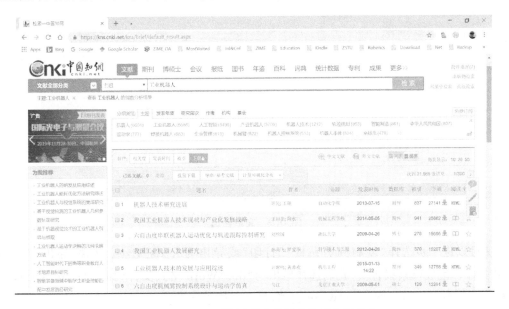

图 D.7　"工业机器人"检索结果按下载次数排序

接下来，我们把检索结果保存下来。

⑧ 勾选第 1 条到第 5 条文献，如图 D.8 所示。

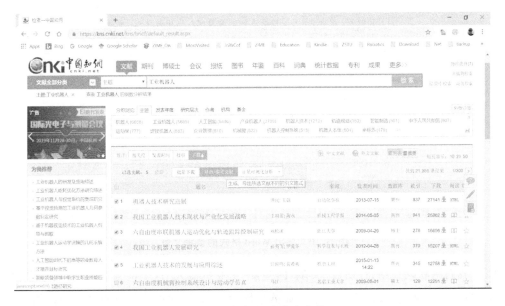

图 D.8　生成、导出参考文献

⑨ 单击"导出/参考文献"，将生成、导出所选文献不同的引文格式，如图 D.9 所示。

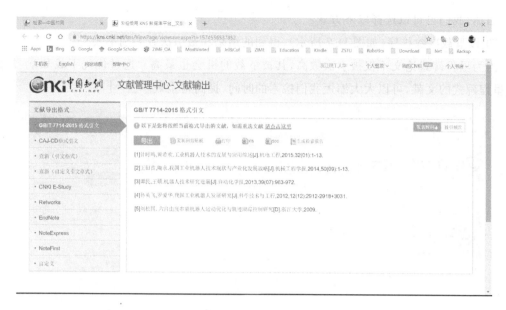

图 D.9 文献输出

⑩ 可以进一步选择"复制到剪贴板""打印"或导出为".xls"Excel 文件、导出为".doc" Word 文件、"生成检索报告"。我们选择导出为"doc"Word 格式文件(如图 D.10 所示),生成了"CNKI-637101822172203750"Word 文档,保存了检索结果。

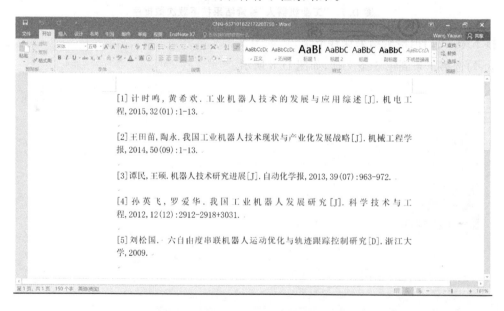

图 D.10 导出 Word 格式的参考文献

刚才我们用默认选项,在所有"文献"内检索"工业机器人"这个检索词,得到的结果包含期刊论文、博硕士论文、会议论文等结果。如欲精炼检索结果,比如说搜索有关工业机器人的"博硕士论文",只要在页面顶部选择"博硕士"即可,结果如图 D.11 所示。还可以选择"百科",结果如图 D.12 所示;或检索"专利",结果如图 D.13 所示。

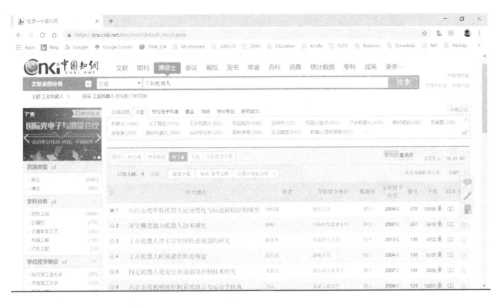

图 D.11　搜索博硕士论文库

图 D.12　搜索百科

我们还可以按照以下步骤进行高级检索：
① 单击检索框右侧的"高级检索"。
② 在主题检索框内输入"定位估计"。
③ 在作者检索框内输入"王耀军"。
④ 单击"检索"键。

得到作者王耀军、林勇刚在《计算机工程与引用》发表的《压缩感知下的自适应生源定位估计》论文，如图 D.14 所示。

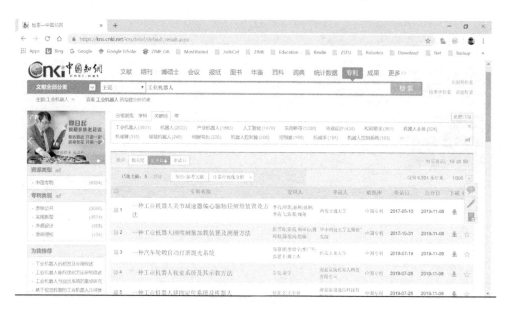

图 D.13　搜索专利

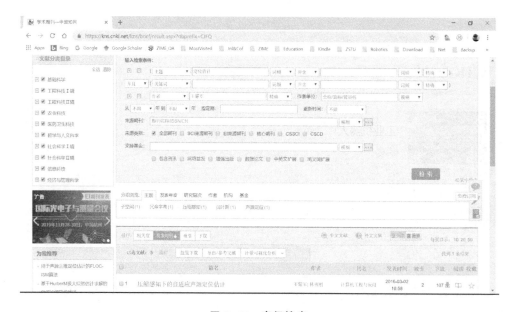

图 D.14　高级搜索

除了"高级检索",还有"专业检索""作者发文检索""句子检索"等检索方法,有兴趣的读者可以分别尝试各种检索方法,找到最适合自己习惯的方法使用。最基本的检索手段就是进入首页显示的"一框式检索"。如何选择检索方式,要根据检索需求、熟悉程度和个人习惯,因人而异。

中国知网 CNKI 还提供了检索结果可视化分析工具,相当于有一个机器人在帮助阅读,从而提高文献阅读和分析的速度。

① 如图 D.14 所示,文献清单上部单击"计量可视化分析"菜单栏。

② 在"已选文献分析"或"全部检索结果分析"下拉式菜单中选后者。分析结果如图 D.15 所示。

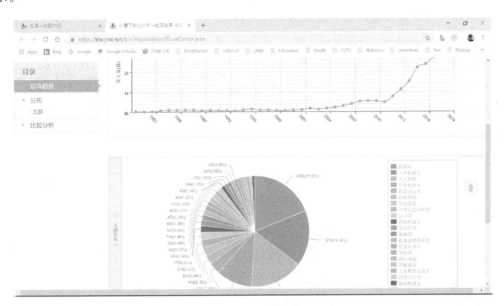

图 D.15 "工业机器人"检索结果分析

除了中国知网，常用的中文数据库还包括：
- 万方数据 http://www.wanfangdata.com.cn/。
- 维普咨询 http://qikan.cqvip.com。

这两种库的使用方法和中国知网大同小异，掌握了一种库的检索方法，同学们可自行尝试使用。

D.3 Retrieval of English database：ISI Web of Knowledge

英语是国际化语言，全世界使用国家和使用人数最多。因此，很多数据库都是英文的。目前国内常用的英文数据库都整合在大学图书馆的电子资源中。常用的英文数据库有 ISI Web of Knowledge、Science Direct、Ei Village2、Springer Link、IEEE/IET Electronic Library (IEL)等。

接下来我们一起来学习如何使用 ISI Web of Knowledge。在 ISI Web of Knowledge 平台上，Web of Science 影响最大、使用最多。Web of Science 是英文检索的文摘数据库的集合，其文献记录来源于 11 000 多种学术期刊及国际会议。国际上一致认为 Web of Science 收录的期刊是核心期刊，认可其学术水平，其中工程领域的科学引文索引（Science Citation Index，SCI）收录的文章，是学术界公认的反映较高学术水平的标志，简称为 SCI 收录。

当然，根据学术水平和引用水平，中科院有一个 SCI 收录一区到四区分类，一区的都是各个领域内的顶级刊物。当然也有一些期刊（如《Mechanism and Machine Theory(MMT)》）虽委身二区，但在其专业领域内是名副其实的权威刊物。

ISI Web of Knowledge 一般通过大学 IP 段登录，其链接在各个大学的图书馆电子资源页面。大学校园内的 IP 一般都可以访问该学校订阅的 ISI Web of Knowledge 资源，且可下载全文。

① 如在浙江理工大学图书馆电子资源首页找到常用数据库链接，如图 D.16 所示。

② 单击 Web of Science 右侧的"官网"就可以进入 ISI Web of Knowledge 网站首页，如图 D.17 所示。

图 D.16　浙江理工大学图书馆电子资源

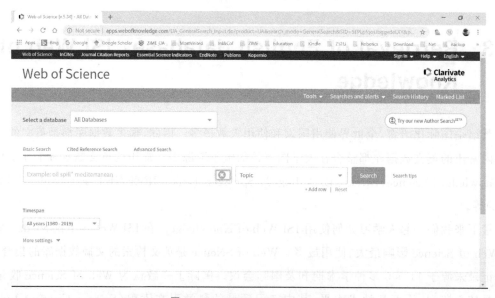

图 D.17　ISI Web of Knowledge

我们看到了经典的 ISI Web of Knowledge 全英文界面。对使用英语语言有困难的读者,还有以下两种方式可选:

③ 请单击页面右上角"English"按钮。

④ 选择下拉式菜单中的"简体中文",则页面显示全中文界面,如图 D.18 所示。

既然我们选择学习专业英语,学习英文数据库的使用,这里强烈推荐大家优先选用默认语言"英文"检索,因此接下来所有图例我们均使用英语界面。

图 D.18　切换语言到中文

⑤ 页面左上角选择数据库(Select a database)选 Web of Science 核心合集(Web of Science Core Collection)。

⑥ 按默认"主题(Topic)"搜索。

⑦ 在搜索框内键入"Kinematic analysis and optimum design of a novel 2pur-2rpu parallel robot",如图 D.19 所示。

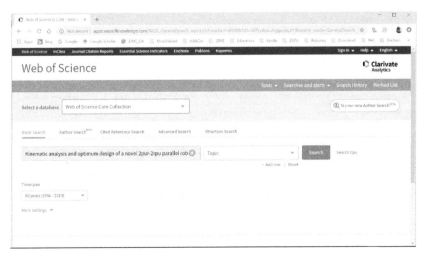

图 D.19　搜索某个文献

⑧ 单击"Search（搜索）"按钮，得到如图 D.20 所示页面。检索到笔者于 2019 年发表在《Mechanism and Machine Theory》的文献。

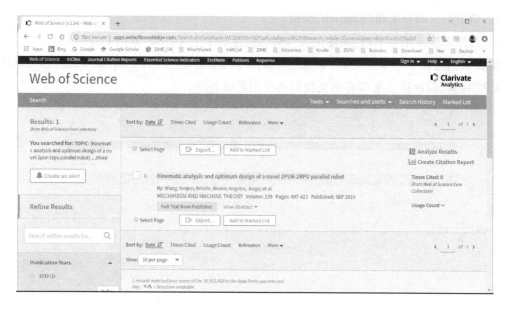

图 D.20　搜索结果

⑨ 单击文献标题，进入文献详细信息页，如图 D.21 所示。在论文详细信息页面中，可看到作者及机构信息、期刊信息、摘要、关键词、引文信息等。

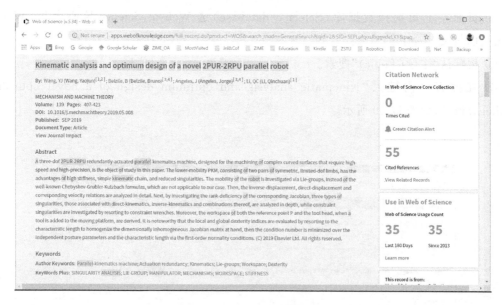

图 D.21　文章详细页面

D.4 Other resources

D.4.1 Bing Academic

除了在大学 IP 内使用学校订购的英文电子资源库搜索文献，我们还可以使用微软的"必应学术"来搜索需要的文献。在浏览器地址栏键入"https://cn.bing.com/academic"即可进入必应学术，如图 D.22 所示。搜索笔者的文献"Kinematic analysis and optimum design of a novel 2pur-2rpu parallel robot"，结果显示如图 D.23 所示。单击文章标题，进入如图 D.24 所示文章详细页面。如需引用该参考文章，则单击"引用(Cite)"，显示如图 D.25 所示。

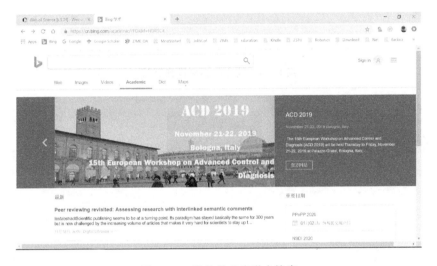

图 D.22 微软的必应学术搜索

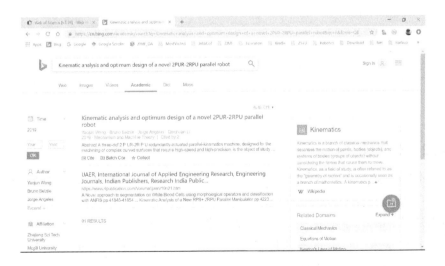

图 D.23 论文的必应搜索结果

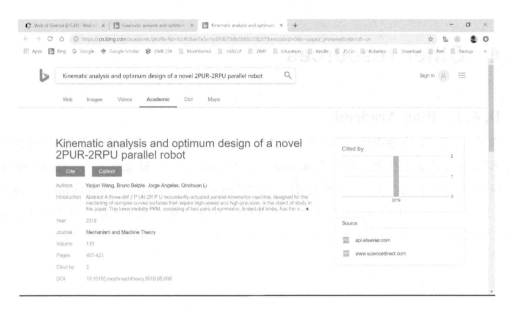

图 D.24　论文的必应详细页面

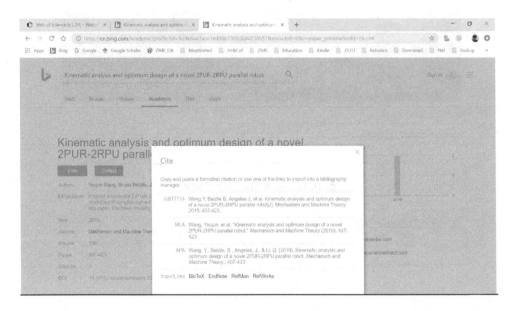

图 D.25　必应引用文献

D.4.2　Google Scholar

谷歌学术(Google Scholar)是国外使用率最高的文献搜索引擎,不过由于未能履行中国的法律和法规,在国内一般无法使用 Google 的服务。据说在一些学校内,可以登录并使用谷歌学术。特别是到了国外,学术界一般都用谷歌学术搜索文献,为了全面了解搜索引擎,下面我们简要介绍谷歌学术。在浏览器地址栏键入"https://scholar.google.com"即可进入,如图 D.26 所示。同样搜索笔者的文献"Kinematic analysis and optimum design of a novel 2pur-2rpu parallel robot",结果显示如图 D.27 所示。单击文章标题,直接跳到 Science Direct

网站的文章详细页面如图 D.28 所示，能够方便获取全文。可以发现，对于英文文献，谷歌学术搜索是方便快捷且广受用户认可的检索工具。

图 D.26　谷歌学术搜索

图 D.27　论文的谷歌搜索结果

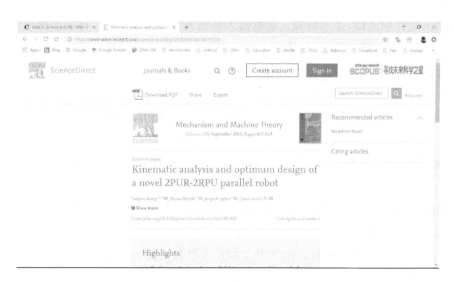

图 D.28　论文的 Science Direct 详细页面

D.4.3 New ways of information retrieval

1. CNKI global academic express

手机日益普及,基本人手一台。那么,我们能否用手机来获得一些信息,提高工作效率呢?让我们一起来了解手机App"全球学术快报",它是CNKI推出的"全球文献整合,学术热点速递"工具。使用方法如下:打开手机→进入手机应用商店→应用商店内搜索"CNKI"→搜索结果选择"CNKI全球学术快报"→再按步骤下载和安装即可。也可以用手机扫描二维码(见图D.29)获取

图 D.29　CNKI 全球学术快报二维码

下载。App 的使用比较简单,同学们可以根据提示使用。还有一个笔者经常用的 App 是"超星阅读器",下载电子图书后可离线观看,利用上下班通勤时间,笔者看完了好多平时没时间看的书。

2. TED

We don't build the lives we want by saving time.

We build the lives we want,

and then times saves itself.

There is time.

Even if you are busy,

we have time for what matters.

And when we focus on what matters,

we can build the lives we want in the time we've got.

耳畔又传来 Laura 铿锵有力的演讲声,即使再忙,时间总是有的,做自己认为重要的事情,过自己想要的生活。

在我刚决定在职读博,深入研究工业机器人系统及应用的日子里,有很多次想放弃:工作实在太忙,家里老小需要照顾,如何抽出时间?年底跟在美国的同学通话诉苦,他建议我看一下 TED,特别是 Laura Vanderkam 的新演讲"How to Gain Control of Your Free Time(如何掌控你的自由时间)"。那是我时间管理的转折点。

TED 是美国一家非营利机构,由 Technology(科技)、Entertainment(娱乐)及 Design(设计)三个英文单词首字母组成。它以组织 TED 大会而著称,以"Ideas Worth Spreading(值得传播的创意)"为宗旨。TED 演讲者要么是某一领域的佼佼者,要么是某一新兴领域的开创人,讲述自己非同寻常的经历和故事。

TED 国际会议于 1984 年第一次召开,由里查德·沃曼和哈里·马克思共同创办,尝试用思想的力量来改变世界。通过观看 TED 演讲视频,不仅可以获取新的信息和创意、令人脑洞大开,还可以提高英语水平,推荐大家多看 TED,学习英语的同时提升精气神。TED 的获取途径主要有以下 3 种:

● TED 网页:https://www.ted.com。

- TED 网易公开课:https://open.163.com/ted/(带中文字幕,但只有部分 TED 演讲)。
- 手机 App:搜索"TED"按提示下载、安装、使用即可。

TED 页面如图 D.30 所示。在搜索框输入"Robot",显示关于机器人的 TED 演讲,如图 D.31 所示。选择第三个"A human-robot dance duet(人机共舞)"即可进入观看页面,如图 D.32 所示。

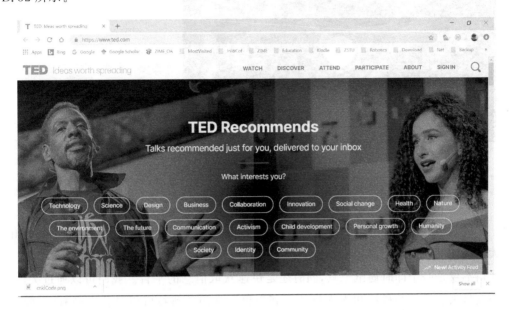

图 D.30　TED

图 D.31　有关机器人的 TED 演讲

3. WeChat

很难想象:失去微信,人类将会怎样？与亲友联系,网上购物,线上学习……很多人早上起

Introduction to Robots and Robotics ◎ 机器人专业英语

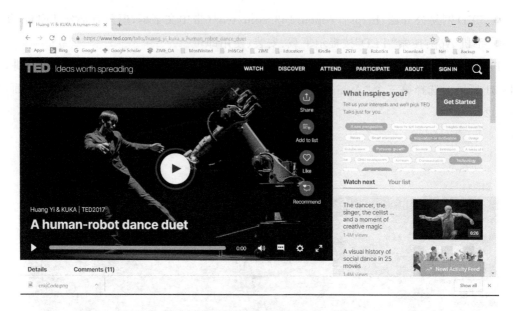

图 D.32　机器人与人共舞（Huang Yi and KUKA）

来第一件事是"刷微信"，晚上睡觉前最后一件事还是"刷微信"。微信把我们的时间碎片化了，那么大家都如何利用这时间碎片？

　　微信不仅是一个即时通讯工具，同时还是快速获取信息的平台。我们可以通过关注一些微信公众号来获取一些科技和人文类的最新资讯。常用的科技资讯类公众号有"虎嗅 App""极客公园 Geekpark""36 氪""果壳网""掌上科技馆"等，大家可以选择自己喜欢的关注，利用闲暇时段、碎片时间获取科技资讯。"极客公园 Geekpark"等公众号界面如图 D.33 所示。大家也可以关注"重燃阅读"公众号来获得最新原版英语资料，学习英语并获取信息。

图 D.33　微信科技资讯公众号

Appendix E

Tell & Show

E.1 Chapter 1

1. Group discussion and self-introduction.
2. What is a robot?
3. What cannot a robot do?

E.2 Chapter 2

1. Share your experience with robots.
2. Tell your view on what a robot should do and should not do, from a viewpoint of the human being.
3. How many types of robots have your learned in this chapter?

E.3 Chapter 3

1. What is a actuator, its definition?
2. What can a actuator do?
3. Tell us all the actuators that you know, explain in more detail the application of these actuators.

E.4 Chapter 4

1. What is a sensor?

2. In what means, does a sensor in a robot function as to those of eyes, ears, nose, mouth, and skin?

3. What sensors should a robot have, if they are to perform like a human being?

E.5 Chapter 5

1. How to control a robot?

2. How many controllers are available in the field of robotics?

3. Do you know any programming languages of a typical industrial robot?

E.6 Chapter 6

1. Show some robot wrists, hands, and grippers to your group.

2. Imagine what a robot can do, if its wrist, hand, or gripper is cut off, tell and show the result.

3. Can you design a new gripper for a specialized manipulation?

E.7 Chapter 7

1. Discuss your understanding of robot kinematics?

2. What do you think about robot dynamics?

3. Tell the differences between robot kinematics and dynamics.

E.8 Chapter 8

1. Tell and show typical performance specification of industrial robot.

2. Is a robot the bigger the better, why?

3. Look up a dictionary, find the meaning of the word 'performance'.

E.9　Chapter 9

1. Use the available search engine, to find the most famous industrial-robot companies in the world, and in China too.

2. Tell the differences between a industrial robot and a humanoid robot.

3. Do you know what is parallel robot, compared to serial robot?

E.10　Chapter 10

1. Other than those mentioned in the text, can you tell more ethics concerns as a college student?

2. Safety, or performance, which is the most important element when design, manufacture, apply robots?

3. Tell your understanding of Asimov's Three Laws of Robotics.

E.11　Chapter 11

1. Will your skills substitute by a robot?

2. Do you think, in the near future, can a robot be as smart as a human being?

3. In order not to be substituted by a robot, what skills should you acquire at school? Share your opinions with other groups in the class.

Appendix F
World of robotics

欢迎来到机器人大观园！所有关于机器人的国内外知名的图书、课程、视频、论坛、网站等，都能在这里找到，这是一个属于机器人的世界。

F.1 Classic books on robots and robotics

1. J. J. Craig. Introduction to Robotics: Mechanics and Control[M]. 3rd Edition. New Jersey: Person Prentice Hall, 2005.

2. L. W. Tsai. Robot Analysis: The Mechanics of Serial and Parallel Manipulators[M]. New York: John Wiley & Sons, 1999.

3. B. Siciliano, O. Khatib. Springer Handbook of Robotics[M]. Berlin: Springer, 2016.

4. J. Angeles. Fundamentals of Robotic Mechanical Systems: Theory, Methods, and Algorithms[M]. 4th Edition. Heidelberg: Springer, 2015.

5. L. Sciavicco, B. Siciliano. Modelling and Control of Robot Manipulators[M]. 2nd Edition. London: Springer-Verlag, 2000.

6. S. Y. Nof. Handbook of Industrial Robotics [M]. 2nd Edition. New York: Wiley, 1999.

7. Y. Koren. Robotics for Engineers, McGraw-Hill, New York, 1985.

8. B. Siciliano, L. Sciavicco, L. Villani, et al. Robotics: Modelling, Planning and Control[M]. London: Springer-Verlag, 2010.

9. T. Bajd, M. Mihelj, M. Munih. Introduction to Robotics[M]. London: Springer, 2013.

10. A. Winfield. Robotics: A Very Short Introduction, Oxford University Press, Oxford, 2012.

11. R. P. Paul, Robot Manipulators, MIT Press, Cambridge, MA, 1981.

12. John J. Craig. 机器人学导论[M]. 负超,等译. 3版. 北京:机械工业出版社,2005.

13. 熊有伦. 机器人学[M]. 北京:机械工业出版社,1993.

14. 蔡自兴. 机器人学[M]. 北京:清华大学出版社,2000.

15. 日本机器人学会. 机器人技术手册[M]. 北京:科学出版社,1996.

F.2　Journals and magazines on robots and robotics

1. Robotics World.
2. IEEE Transactions on Robotics and Automation.
3. International Journal of Robotics Research (MIT Press).
4. International Journal of Robotics & Automation (IASTED).
5. 机器人.
6. 智能机器人.
7. 机器人技术与应用.
8. 机器人产业.

F.3　Websites on robots and robotics

1. IEEE Spectrum:
   https://spectrum.ieee.org/blog/automaton.
2. International Federation of Robotics:
   https://ifr.org.
3. RIA-Robotics Online-Industrial Robotics:
   https://www.robotics.org.
4. Robohub:
   https://robohub.org.
5. The Robot Report:
   https://www.therobotreport.com.
6. Phys.org-Robotics:
   https://phys.org/technology-news/robotics.
7. MIT Technology Review-Robotics:
   https://www.technologyreview.com/c/robotics.
8. KUKA Robotics:
   https://www.kuka.com.
9. ABB Robotics:

https://new.abb.com/products/robotics.

10. Fanuc Global:

    https://www.fanuc.com.

11. Yaskawa Global:

    https://www.yaskawa-global.com/product/robotics.

12. Yaskawa Motoman:

    https://ww.motoman.com.

13. Google AI Robotics:

    https://ai.google/research/teams/brain/robotics.

14. Robotic Mechanical Systems Laboratory:

    http://www.cim.mcgill.ca/rmsl/Index/index.htm.

15. Modern Robotics:

    http://hades.mech.northwestern.edu/index.php/Modern-Robotics.

F.4 Robotics open courses

Future Learn

1. Begin Robotics:

   https://www.futurelearn.com/courses/begin-robotics.

2. Introducing Robotics-Robotics and Society:

   https://www.futurelearn.com/courses/robotics-and-society.

3. Introducing Robotics:

   https://www.futurelearn.com/programs/robotics.

edX

1. Robotics:

   https://www.edx.org/course/robotics-2.

2. Robotics Foundations I-Robot Modeling:

   https://www.edx.org/course/robotics-foundations-i-robot-modeling.

3. Autonomous Mobile Robots:

   https://www.edx.org/course/autonomous-mobile-robots.

Coursera

1. Robotics Specialization:

   https://www.coursera.org/specializations/robotics.

2. Modern Robotics, Course 1: Foundation of Robot Motion:

https://www.coursera.org/learn/modernrobotics-course1.

3. Building Arduino Robots and Devices:
https://www.coursera.org/learn/arduino.

Udemy

1. Electricity & Electronics-Robotics, learn by building:
https://www.udemy.com/course/analog-electronics-robotics-learn-by-building.

2. Tech Explorations Arduino Robotics with the mBot:
https://www.udemy.com/course/arduino-robotics-with-the-mbot.

3. Industrial Robots:
https://www.udemy.com/course/industrial-robotics.

Udacity

1. 机器人开发(英):
https://cn.udacity.com/course/robotics-software-engineer-nd209-cn.

2. 机器人人工智能:
https://cn.udacity.com/course/artificial-intelligence-for-robotics-cs373.

Others

1. Stanford Online Course-Introduction to Robotics:
https://online.stanford.edu/courses/cs223a-introduction-robotics.

2. 网易公开课:机器人技术与应用:
http://open.163.com/newview/movie/
courseintro? newurl=％2Fspecial％2Fcuvocw％2Fjiqirenjishu.html.

3. 网易公开课:机器人学:
https://open.163.com/newview/movie/free? pid=MCQSVOF2A & mid=MCQT097J9.

F.5 Other resources

1. 全国工业机器人技术应用技能大赛:
https://www.miiteec.org.cn/plus/list.php? tid=28.

2. 世界技能大赛(WorldSkills)-Mobile robotics:
https://worldskills.org/skills/id/9/.

3. TED 机器人主题演讲:
https://www.ted.com.(搜索 Robotics)

Appendix G

Words and expressions

abdomen	腹部
absolute	绝对的
absurd	荒诞的
accelerate，acceleration，accelerometer	加速，加速度，加速度计
accuracy	准确性，准确度
acknowledge	确认
actroid	类固醇
actuate，actuation，actuator	开动；驱动；驱动器、执行器
additionally	另外
advantage	优点
agency，agent	代理，媒介；介质，试剂，代理人
albeit	尽管，虽然
algorithm	算法
alleviate	减轻，使缓和
alternatively	或者，二选一
ambient	周围的，包围物
ambiguously	模棱两可的
amplify，amlifier，amplitude	放大；放大器；振幅
analogue	模拟量
analytical	分析的
android	机器人
angular	角度的，角形的
animation	动画
anthropomorphic	拟人的
appearance	外表，表象
application	应用
appropriate	适当的，合适的

approximation	近似
arithmetical	算术的
arrangement	安排
articulate	环接,连接
artifact	人工制品,加工品
artificial, artificial intelligence	人工的;人工智能
assembly line	流水线
assistant, assistance	助手;协助
associated	关联的
assumption	假设
assure	保证
asteroid	小行星
attachment	附件
attribute	属性
autonomous, autonomy	自治的;自治
autopilot	自动驾驶
auxiliary	辅助的
avoidance	回避,避免
axis	轴
beacon	信标,标示
behavior	行为
bewild	迷惑
bimetallic	bi-前缀代表双、两个,双金属的
binary	二元的,二进制的
biochemistry	bio-chemistry,生物化学
biology	生物学
blur	模糊
boring	无聊的,或钻探、钻孔
breakthrough	突破
bulk	块,批量
bump	颠簸,碰撞
butler	管家
calculation	计算
cam	凸轮
capability	能力

capacity	容量
capillary	毛细血管
Cartesian	直角坐标
cartoon	动画片
catastrophic	灾难性的
certification	认证,认证书
characteristic	特性
choreograph	编舞
chronological	按年代顺序排列的,先后的
circulate	流通,循环
circumstance	环境
classification	分类
clearance	间隙
cognitive	认知的
coin	硬币,这里指创造一个新词
collision	碰撞
combination,combinational	组合;组合的
comet	彗星
companion	伴侣,同伴
comparison	比较
compensator	补偿器
compile	编译
complaint	抱怨
complex	复杂的
compliance	合规,这里指柔顺
comply	遵守
component	零件,元件,组件
compound	复合
compress,compressor,compressibility	压缩;压缩器;可压缩性
computer-aided design	计算机辅助设计
comtemplate	思索,考虑
conductor	导体
configuration	组态,配置
connectivity,connector	连通性;连接器
conscience	良心,良知

considerable	相当的,重要的,可观的
consideration	考虑
consistency	一致性
console	控制台,操纵台
constraint	约束,抑制
construction	施工
contender	竞争者
contribution	贡献
controller	控制器
controversial, controversially	有争议的;有争议地
convention	惯例
converse, conversation, conversationally	交谈,相反的;对话;对话地
coordinate, coordinate system	坐标;坐标系
corrosion	腐蚀
cosmetics	化妆品
counterpart	对方,另一方
crucial	关键的
crude	粗糙的,天然的,未成熟的
cylinder, cylindrical	圆柱,圆柱体;圆柱形的,气缸的
damp	阻尼,衰减
dangerous	危险的
debug	调试,排故
decelerate	减速
deceptive	欺骗性的
decontaminate	净化,去污
definition	定义
deflect	偏转,转向
deformation	变形,畸变
degree of freedom	自由度
delay	延迟
deliberately	故意地
deliver	交付
demonstrate	演示
dependable, dependability	可依靠的;可靠性
derivative	导数

destination	目的地
detachment	分离
detection	检测
determinant	决定因素,行列式
detonate	爆炸,爆破
dexterity	灵巧性
differential	微分的,差动齿轮
dimension	尺寸
discourse	谈论
discretion	判断,慎重
displacement	位移
disposal	处置
distinction	区别,差别
distinguish	区分
diverting	转移
dominant	优势的,主流的
dramatist	剧作家
drill	钻头
driverless	无人驾驶
droid	机器人
drone	无人驾驶飞机
dull	呆板的
dust	灰尘
dynamics	动力学
elaborate	阐述
elasticity	弹性
electric, electricity	电的;电,电学
electromagnetic	电磁的
electromechanical	机电的
electrostatic	静电的
embodiment	体现,化身,具体化
emergency	紧急情况
emitter	发射器
emotional	情感的
empathy	移情作用

employ, employee, employer, employment	雇用;雇员;雇主;就业
enclosure	外壳,封装,包围
encoder	编码器
end-effector	末端执行者
endpoint	终点
end-stop	终点挡板
endurance	耐力
energetically	大力地,有力地
equivalent	相当的,差不多的
erratic	不稳定的
essentially	实质上的
estimate	估计
ethics	伦理
exceptional	例外的,特别的,异常的
exobiology, exobiologist	太空生物学;太空生物学家
explicitely	明确地
exploration	勘探
external	外部的
exteroceptive	外感受性的
extraordinary	非凡
extremely	非常的,极端的
fabricate	制造
facility	设施
failsafe	故障安全
fantasy	幻想
fatigue	疲劳
fault	故障
feasible	可行的
feedback	反馈
fetch	取
fidelity	保真度
finalize	使结束,定稿
fixture	夹具,固定挡销
flange	法兰
flexible, flexibility	灵活的;灵活性

flora and fauna	植物群和动物群
fluctuation	波动,振荡
footprint	轨迹,足迹
forcep	镊子
foreseeable	可预见的
forklift	叉车
formulation	公式
fortunately	幸好
fractional	小数的,分数的
fraught	充满的,含有的
fume	烟气,蒸气
functionality	功能性
fundamental	基本的,核心的
geminoid	双子星
geology, geologist	地质学;地质学家
geometric(同 geometrical)	几何的,几何学的
geometry, geometrically	几何学;几何学地
gesture	手势,姿态
gripper	抓手
guidance	导引装置,引导
handle	处理
hazard, hazardous	危险因素;危险的
helical	螺旋形的
hinge	合页
hobby	爱好
homegeneous	相同的,相似的;均匀的,单一的
humanity	人性,人类
humanoid	人形,类人
hybrid	杂种,杂交
hydraulic	液压的
hydrostatic	静水力学,流体静力学
hydrothermal	热液的
icon	图标,图像
identification	鉴定,识别,辨认
imitation	模仿

immersive	沉浸式的
implausible	难以置信的,不真实的
impressive	给人印象深刻的,感人的
improvement	改善
in conjunction	一并
inaction	无所作为,不作为
inadequate	不足
incompletely	不完全
incorporate	合并,混合
incremental	增加的,增量的
industrial	产业的
inertia	惯性
infrared	红外线的
inherent	固有的
initiate and terminate	发起和终止
initiation	引发,发起
inspection	检查
inspirational	鼓舞人心的
installation	安装,装置
instrument	仪器
insulation	绝缘
intelligent	智能的
interaction	相互作用
interchangeably	可互换的
interestingly	有趣的是
interference	干扰,干涉
interlock	互锁
intermediate	中间的
internal	内部的
interpretation	解释
interrupt	打断,中断
intervention	介入
intimately	亲密地
intimidating	吓人的
invasive	侵入性的

investigate	调查
irrelevant	不相关的
isolation	隔离
Jacobian	雅可比矩阵
jig and clamp	夹具,夹子
jog	步进
joint	关节,接头
joystick	操纵杆,游戏杆
judgement	判断
keyboard	键盘
keyhole	锁孔,钥匙孔,键孔
keyword	关键词
kinematics	运动学
labor-intensive	劳动密集型的
landmine	地雷
laparoscopic	腹腔镜的,内窥镜的
laptop	笔记本电脑
lead-through teaching	引导式示教,机器人示教的一种模式
linearity	线性度
linguistically	语言上地
linkage	连杆
literature	文献
longitudinal	纵向地
loop, open-loop, closed loop	回路;开环;闭环
lurk	潜伏
machinery	机器
magnetostriction	磁力控制、磁弹性
magnitude	大小,强度,等级
maintainer	维修工,养护工
maintenance	保养,维护
maneuver	机动,演习,策略
manipulate, manipulator, manipulation	操纵;机械手,操纵器;操纵
manufacture, manufacturer	生产,制造;生产商,制造商
mass	物体的质量
measurement	测量

mechanical	机械的
membrane	膜
microprocessor, microscopic, microsurgery	微处理器；显微的；显微外科
milestone	里程碑
miniature, miniaturization	微型的；小型化，微型化
minicomputer	微型机
minimally	最低限度地
miscellaneous	杂的，各种的，多方面的
mistress	女主人
model	模型
modesty	谦虚
modular	模块化的
moral	道德
mount	安装
multifunctional	多功能的
munition	弹药
muscle	肌肉
navigate, navigation, navigational	导航，航行；导航，航行；导航的，航行的
noteworthy	值得注意的
objective	目的
obstacle	障碍，障碍物
offshore	离岸的
operator	运算子
optical	光学的
orbit	轨道
ordnance	军械
orientate, orientation	定位，定向；定位，定向
outline	大纲
overlap	交叠
overload	超载
overshoot and undershoot	过冲和下冲
palletize	堆垛
parallel robot	并联机器人
parallelogram	平行四边形
parameter	参数

particularly	尤其
payload	有效载荷
pendant teaching	示教器示教,机器人示教的一种模式
perception	知觉,感知
performance	性能
perish	毁灭,死亡
perpendicular	垂直的
personify, personnel	拟人化;人员,职员
perspex	有机玻璃
pheriphery	周边
photocell	光电管
photoelectric encoder	光电编码器
pick-and-place	取放
piezoelectricity	压电
piston	活塞
planetary, planetoid	行星的;小行星
pneumatic	气动
polymer	聚合物
polynomial	多项式
popular, popularity, popularize	流行的;名声,流行;宣传,普及
porter	搬运工
portray	描写
possibility	可能性
posture	姿态
potentially	潜在地
potentiometer	电位计
precision	精确,精度
predetermined	先前已决定的,业已决定的
predictable	可预测的
premise	提论,假定
prismatic	移动的
privacy	隐私
productivity	生产率
programme, programmability	程序;可编程性
prominence	突出

Words and expressions

propensity	倾向,习性,爱好
property	属性
proportional	成比例的
prostate	前列腺
prototype	原型
proximity	接近
quartz	石英
radiation	辐射
ram	滑枕
realistically	现实地
receiver	接收器
recharge	充电
recognize, recognition	承认;认可
reconnaissance	勘测,探测
reducer	减速器
reduction gearbox	减速齿轮箱
redundant	多余的,冗余的
reliable, reliability	可靠的;可靠性
remarkable	杰出的,很多的
remedial	补救的
removal	清除
repeat, repetitive, repeatable, repeatability	重复;重复性的;可重复的;可重复性
replica	复制品
reprogrammable	可重编程的
requisite	必要条件
resistence	阻力,抵抗
resolution	解析度
resonant frequency	共振频率
responsible	负责任的
resultant	结果的
resume	简历
retrieval	检索
revolute	转动的
revolutionize	革新
rigidity	刚性

roadmap	路线图
robot, robotics, roboticist	机器人;机器人学;机器人学家
roll, pitch, yaw	滚动角;俯仰角;偏航角
rotation	回转
roughly	大致
rover	漫游者
safeguard	保障
salvage	海难救助,抢救
sanction	处罚,制裁
sander	砂光机
scalpel	解剖刀
scenario	情节,情景
screw	螺旋,一种数学工具,用于分析机构
seam welding	缝焊
senior	老年人
sense, sensory	感知,感官;知觉的,感觉的
sequence	序列
serial robot	串行机器人
servomotor	伺服马达
shaft	轴
siginificant	重要的,有意义的
simplicity	简单
simulation	模拟
simultaneously	同时地
singular, singularity	奇异的;奇异性
skeleton	骨架
slide	滑动
small batch	小批量
smooth	光滑的
socialization	社会化
sonar	声呐
sophisticated	久经世故的,诡辩的
spare	备品、备件
spatial	空间的
specification	规格

spherical	球形的
spot welding	点焊
spray painting	喷漆
stability, stabilization	稳定性；稳定
standardization	标准化
stepper motor	步进电机
stiffness	刚性，刚度
strain gauge	应变仪
stressful	压力重的，受力大的
stroke	行程
submarine	潜艇
submersible	潜水的
subroutine	子程序
substitution	替代
sufficiently	充分地
surgeon, surgical, surgery	外科医生；外科的；外科手术
surveillance	监视
sustainable, sustainability	可持续的；可持续性
suture	缝线
swarm	一群
swept	延伸，伸展
symbolic	象征性的
tachometer	转速计
teach	示教（机器人）
technical	技术的
tele-haptics, telemanipulator, tele-operated	远程触觉；远程操纵器；遥控
tentative	试行的，试验性的
tether	系绳
textbook	教科书
therapeutic	治疗的
toggle	拨动
torque	扭力
tow	拖
trade-off	权衡，折中
traditionally	传统上

trajectory	轨迹
transaction	交易,执行,办理
transducer	传感器,变换器
translation	翻译
transmit, transmission	传送;送达
transparent	透明的
transport	交通运输
trap	陷阱
tremble	颤抖,震动
trunk	树干,主干
trustworth	信任度
twist and wrench	运动螺旋和力螺旋
ultimately	最终
ultrasonic	超音波
unauthorize	取消授权,未授权
underpin	支持,支撑
undertake	承担
underwater	水下的
universal	普遍的
unpredictable, unpredictability	不可预测;不可预测性
unsettle	不安
unwittingly	不知不觉地,无意识的
vague	模糊
valet	仆役,随从
vane	叶片,风环
variation	变异
varied	变化的,不同的
vascular	血管的
velocity	速度
vendor	供应商
vent	出口,出孔,通路
vibration, vibrational	振动;振动的
vice versa	反之亦然
voltage	电压
vulnerable, vulnerability	易损坏的;易损性

walk-through teaching	演练示教,机器人的一种示教模式
warehouse	仓库
wear and tear	磨损,消耗,耗损
whisker	晶须
widespread	普遍的,分布广泛的
wit	机智
workbench	工作台
workforce	劳动力
working volume	工作空间、工作范围
workpiece	工件
workspace	工作区
workstation	工作站
wrist	腕
zoomorphic	兽形的

Reference

[1] 王卫平. 英语科技文献的语言特点与翻译[M]. 上海:上海交通大学出版社,2009.

[2] 严俊仁. 900科技英语长句难句分析和翻译[M]. 北京:国防工业出版社,2010.

[3] Jorge Angeles. Fundamentals of Robotic Mechanical Systems: Theory, Methods, and Algorithms(volume 124) [M]. [S. l.]:Springer Science & Business Media,2013.

[4] John J Craig. Introduction to Robotics: Mechanics and Control(volume 3)[M]. NJ: Pearson/Prentice Hall Upper Saddle River,2005.

[5] D. McCloy, D. M. J. Hariis. Robotics: An Introduction[M]. [S. l.]:Springer Science & Business Media,1986.

[6] B. Siciliano, O. Khatib. Springer Handbook of Robotics[M]. [S. l.]:Springer Science & Business Media,2016.

[7] David Todd. Fundamentals of Robot Technology: An Introduction to Industrial Robots, Teleoperators and Robot Vehicles [M]. [S. l.]: Springer Science & Business Media,2012.

[8] Alan Winfiele. Robotics: A Very Short Introduction[M]. [S. l.]:OUP Oxford,2012.

Acknowledgement

首先,我要感谢浙江机电职业技术学院的各位领导、同事以及我的学生,正是有了你们的支持,我才得以来到枫叶之国——加拿大脱产学习一年,静下心来编写本书。

其次,特别感谢国家留学基金委、浙江省教育厅的资助,让我安心学习而不必操心生计。

言语之情无法表达对 Jorge Angeles 教授的感激,讲义编写过程中,教授手把手指导、指正内容、修正拼写和语法及表达错误,本书得以顺利完工完全依赖教授的全程指导。

感谢家人:母亲的教诲,妻子的默默支持,还有女儿的期望——期望爸爸成为她学习的榜样,这些一直鼓舞着我坚持到底。

感谢亲爱的同学们:你们对学习的渴望、对未来的展望、对前途的期望是我编写本书的原始动力。